# Plant Nutrient Disorders



NIPA® GENX ELECTRONIC RESOURCES & SOLUTIONS P. LTD.
New Delhi-110 034

www.nipaers.com
Browse, Search, Read & Buy...
Print Books
eChapters
eBooks
Publishers
Forthcoming
Subject
Catalogues
New Titles
Online Resources on:
✓ Current Affairs
✓ Reasoning, Logic & Aptitude
✓ Competitive Examinations
✓ English Language Lab
✓ Writing & Pronunciation Tools
✓ Personality Development
✓ Online Programmes for Professionals
✓ Interview Preparation
Pay Using
PhonePe
Google Pay
paytm
PayPal
UPI
CC Avenue

# Plant Nutrient Disorders
## Diagnosis and Management

*Editors*
**A.K. Sarkar**
**P. Mahapatra**

NIPA® GENX ELECTRONIC RESOURCES & SOLUTIONS P. LTD.
New Delhi-110 034

**NIPA® GENX ELECTRONIC RESOURCES & SOLUTIONS P. LTD.**

101,103, Vikas Surya Plaza, CU Block
L.S.C. Market, Pitam Pura, New Delhi-110 034
Ph : +91 11 27341616, 27341717, 27341718
E-mail: newindiapublishingagency@gmail.com
web: www.nipabooks.com

***For customer assistance, please contact***
Phone: + 91-11-27 34 17 17
Fax: + 91-11- 27 34 16 16
E-Mail: feedbacks@nipabooks.com



ISBN: 978-81-19254-80-4



Composed and Designed by NIPA®.

# Preface

Most of us are aware that, there is a continuous decline in the productive potential of natural resources with time. With stepping up of industrial activities and growth in population & meeting its requirements, the per capita availability of land is also decreasing. This, in turn, has put up a challenge to produce more from less land. Therefore, protecting the available land resources from disruptive forces, such as, degradation, loss of fertility, nutrient imbalance, pollution, salinization, acidification, waterlogging etc. is now absolutely essential for future sustainability.

Maintenance of adequate nutrient and water supply in soil is essential for plant growth and metabolism and sustain rich crop harvests. This will ensure food and nutritional security of the growing population and rural prosperity, besides bringing about a positive environmental change, peace and social harmony. What is difficult to understand is that, why such an important issue is not being addressed with the seriousness, that it deserves?

Some of these concerns are: i) a gap of about 10 million tonnes of plant nutrients exists between the amount removed & the amount replenished at the current level of crop production, ii) about 47 million hectare of land under important cropping systems, have deficient available sulphur level, iii) over 30 to 40% area under cultivation in alluvial, black, red and calcareous soil regions have shown Zinc and Boron deficiency in cereals, pulses, oilseeds, vegetable and fruit crops, iv) crop yields in about 10 million hectares of the Indo-Gangetic Plains are plateauing, due to improper and imbalanced nutrient management practices. All these, call for an in-depth analysis, concrete plans and aggressive implementation, taking stake holders into confidence.

This Book on *'Plant Nutrient Disorders: Diagnosis and Management'* is the result of our thought process on the issues raised above. Authors feel that solution lies on a holistic and multi-faceted approach involving soil, water, plant, and environment and their surveillance system as an integral part of the management planning and practices for agricultural development.

I am grateful to the authors involved in the preparation of different chapters in a strict time frame. Their invaluable contribution has been a major driving force in completing the assignment. I am indebted to my wife, Mrs. Sabita for her inspiring attitude, support, patience and encouragement in all such endeavours.

Place : Jadavpur, Kolkata-75. **A. K. Sarkar**

Date : June 2015

# Contents

*Preface* ........ *v*

1. **Introduction** ........ **1**

   *A. K. Sarkar*

2. **Plant Nutrient Disorders** ........ **5**

   *A. K. Sarkar*

3. **Soils of India-Nutrient Status and Management** ........ **27**

   *A. K. Sarkar*

4. **Soil Analysis** ........ **57**

   *P. Mahapatra*

   A. Soil Sampling ........ 57

   B. Instrumental Methods ........ 71

   C. Methods of Analysis ........ 97

   D. Setting up of Soil Testing Laboratory ........ 137

5. **Plant Analysis** ........ **153**

   *S. C. Kotur*

6. **Water Analysis** ........ **177**

   *G. N. Chattopadhaya & Abira Banerjee*

7. **Environmental Monitoring** ........ **215**

   *Dipak Kumar Gupta*

8. **Soil Nutrient Mapping - a case study** ........ **247**

   *A. K. Sahoo, S. K. Singh, Dipak Sarkar & A. K. Sarkar*

9. **Task Ahead** ........ **277**

   *A. K. Sarkar*

## CHAPTER - 1

# Introduction

Soil is the most precious basic natural resource, on which depends the security of food, nutrition, income, biodiversity, livelihood and environment. But, its indiscriminate and unscientific management is a major threat to sustainability of agriculture. Over the years, there is a loss in the productive potential of soils, which is attributed to imbalance in nutrient use, mining of nutrients from soil, poor replenishment, loss of carbon, pollution, decrease in biological activities, waterlogging, soil salinity etc. Thus, land use is not commensurate with land capability.

With increased cropping intensity in major groups of soils , dis-proportionate increase in the rate of nutrient application, the inherent soil fertility is showing a decline with reduction in crop uptake of essential plant nutrients. This has led to lowering of crop yields with deterioration in crop quality with lower value of economic produce. Increased extraction of several nutrients from soil has resulted in multi-nutrient deficiencies, poor efficiency of applied nutrients and lower returns from money spent on fertilisers and other inputs. This process of degradation has to be reversed for the benefit of masses, for whom agriculture is the prime source of livelihood.

Another consequence of this is that, the food (cereals, oilseeds, pulses, vegetables, fruits, coarse cereals etc.) we consume, is of poor nutritional value, if it is grown in soils with poor or low nutrient status. This often leads to widespread mineral deficiency disorders in human beings and animals, such as anaemia, goitre, dental carries etc.

On the above background, the present publication was planned for an analysis of the' Plant Nutrient Disorders', their detection and management concerns.The book cosists of nine chapters. The 2nd. Chapter (plant nutrient disorders) deals with a study of the relationship between soil health and plant health. The role and deficiency symptoms of different essential nutrients for plants have been critically discussed. This chapter also brings into focus, how imbalanced nutrient use, loss of soil microorganisms, and lowering of soil organic matter can change the mineral nutrition of plants. 3rd. Chapter (soils of India-nutrient status & management concerns) discusses issues of land degradation, nutrient mining, current nutrient use trends for crop production, carbon sequestration in soils and their management concerns. The next four chapters (4,5,6,7) focus on the methodologies of detection of nutrient disorders by soil analysis, water analysis, plant analysis and environmental monitoring. The Chapter 'soil analysis' deals with four related components, i.e., soil sampling, instrumental techniques, analytical methods, and ways to set up soil testing laboratories for taking the work forward. The Chapter 5 'plant analysis' provides a detailed scientific information on sampling and analytical techniques and interpretational methods, such as, critical value approach, nutrient concentration, DRIS (diagnosis and recommendation integrated system) and compositional nutrient diagnosis (CND). Water is a primary requirement for intensive agriculture and aquaculture. The Chapter 6 on ' water analysis' critically analyses the methods to judge the suitability of irrigation water for crop production based on parameters like TDS (total dissolved solids), TSS (total soluble salts), BOD (biological oxygen demand), and COD (chemical oxygen demand). Similarly, for freshwater aquaculture, DO (dissolved oxygen), free $CO_2$, pH, alkalinity, Chlorine, ammonia or nitrate concentration are important parameters. Chapter-7 'environmental monitoring', provides interesting details on soil quality, soil pollution and soil resilience. The impact of climate change on agriculture with necessary environmental legislation, standards and monitoring mechanisms have been discussed.

The 8th. Chapter (nutrient mapping - a case study ) provides information, how soil nutrient mapping can help in soil fertility assessment. The soil testing network of the state of Jharkhand is very poor, and therefore, soil nutrient mapping at district and block level can provide a viable alternative to farmer's need of soil testing. A small district-Jamtara for example, was selected for preparation of soil fertility maps. 'Kundahit' block of 'Jamtara' district, was selected for preparation of block level soil nutrient maps. One can understand the importance of such efforts for delineation of nutrient deficient areas.

Chapter 9 (task ahead) provides possible answers to some important questions on the subject. Attempt has been made to provide a stepwise methodology for detection of plant nutrient disorders, their diagnostic techniques, management options, and the resultant benefits. Some of the new areas in this sphere has been identified, which on implementation, can bring about perceptible changes in our current knowledge.

To sum up, the effort is to address the issue of the close inter-relationship among soil-water-plant-environment to raise agricultural productivity by detecting the problems and adopting sustainable management measures for their redressal to realise the potential of the natural resources to achieve food and nutritional security of the rising human population. It is hoped that this will help in societal development, create awareness, and help us march towards a hunger-free world in years ahead.

❑❑❑

## CHAPTER - 2

# Plant Nutrient Disorders

**A K Sarkar**

Plant nutrient disorders result from nutrient imbalances in soils. Nutrients are considered essential for plants, when it satisfies the following criteria (Arnon and Stout,1939):

1. The deficiency of the nutrient element makes it impossible for the plant to complete its life cycle
2. The deficiency of the element is specific and can be corrected only by supply of that element
3. The element is directly involved in the nutrition of the plant

Soil is the major source for most of the essential plant nutrients.Total nutrients reserve in a soil may vary widely due to the different factors of soil formation (parent material, climate, topography, time and biota) and the soil forming processes (laterization, podsolization, humification etc.). Nutrients in the readily available form is the intensity factor, while the ability of the soil to replenish the same, when the available nutrients are in short supply for growing plants,is the capacity factor.

Many soil factors are responsible for the deficiency or toxicity of plant nutrients. These are soil texture, soil pH, organic matter content, calcium carbonate content, soil moisture etc. pH is the master variable, which profoundly influences the availability of nutrients.

Nutrient disorders in acid soils are primarily due to low soil pH and high $Al^{3+}$ concentration in the soil solution and exchange complex. Deficiency of Phosphorus is common due to its fixation in soil and/or reversion to unavailable forms. Poor base saturation results in decreased productivity of such soils. Acid soils are often deficient in Ca, Mg, B, S and Mo. On the other extreme are the saline and alkali soils, characterised by low nutrient ion activities and high ratio of $Na^+/Ca^{2+}$, $Na^+/K^+$, $Ca^{2+}/Mg^{2+}$ and $Cl^-/NO_3^-$ leading to nutrient disorders affecting plant growth.

In waterlogged soils,fall in redox potential as a result of submergence controls the availability of plant nutrients. Nitrogen, sulphur, iron and manganese are important nutrients, which exist in more than one oxidation state and therefore influenced by oxidation-reduction processes in such soils. Desert soils are generally low in clay content (<10%) and are alkaline in nature. Due to coarse texture, soil and water retention is low leading to poor fertility. Majority of such soils are deficient in N, P, K, S and Zn.

For proper plant growth, 17 nutrients are considered essential. These are: carbon(C ), hydrogen (H),oxygen (O), nitrogen (N), phosphorus (P), potassium (K), calcium (Ca), magnesium (Mg), sulphur(S), iron (Fe), zinc (Zn), copper (Cu), manganese (Mn), boron (B), molybdenum (Mo), chlorine (Cl), nickel (Ni). Some of the elements capable of stimulating growth and development even in the presence of adequate supplies of essential elements, are called beneficial or functional (Asher, 1991). Nutrient disorders in plants are caused due to insufficient supplies of essential plant nutrients present in the growth medium or problems in their translocation from soil to plant system mediated by the following factors:

1. Nutrient imbalances in soils. High Ca levels in soils often lead to problems of K or B nutrition. Excess Ca in soil can lead to iron deficiency (lime induced iron chlorosis)
2. Soil related constraints, such as acidity/salinity/alkalinity/low soil moisture often lead to nutrient disorders.
3. Slow rate of release of plant nutrients, such as P, Zn, K in clayey soils from exchange complex to soil solution often result in deficiency of these nutrients.
4. Nutrient interactions, such as P-Zn, Fe-Zn, etc. in soils induce Zn deficiency in soils
5. Improper or no use of micronutrients, such as Zn, B, Fe etc in cultivated soils, in spite of being deficient.
6. Low soil temperature/poor soil moisture/excess soil moisture and other environmental factors also lead to nutrient disorders.

### Soil Health and Plant Health

Plants derive most of the essential nutrients from soil. Thus, plant health depends on the quality of soil. Intensive cropping, imbalanced nutrient use, decline in soil organic matter, non-application of secondary and micronutrients based on soil tests have resulted in a decline in soil health. Long term fertiliser experiments with different cropping systems in some agro-ecological regions reveal that application of N alone (common practice with many farmers) increased soil depletion of available P, thus causing fast appearance of P deficiency symptoms. In plots fertilised with only N, 33 to 129% more soil P was removed as compared to plots receiving no N. The extent of P depletion due to this was highest in alluvial soils and least in red soils. The red soils are generally very poor in available P, and therefore, even a small removal of P is of great significance (Nambiar and Abrol, 1989). Similarly, N and P applied fields result in poor soil available K level and negative K balance (Table 1) under intensive cropping systems at different locations in India.

**Table 1 :** Potassium balance sheet (kg K $ha^{-1}$) for some soils & cropping systems after 7 years

| Cropping system | Treatment | Added (Kg K $ha^{-1}$) | Removed (Kg K $ha^{-1}$) | Balance (Kg K $ha^{-1}$) |
|---|---|---|---|---|
| Ludhiana (Punjab) | NPK | 820 | 1822 | -1002 |
| Maize-wheat-cowpea | NP | 0 | 1260 | -1260 |
| Hyderabad | NPK | 300 | 1196 | -896 |
| Rice-rice | NP | 0 | 1290 | -1290 |
| Palampur (H.P) | NPK | 580 | 930 | -350 |
| Maize- wheat | NP | 0 | 700 | -700 |

Depletion of Potassium from soils was alarming. Even with optimal dose of NPK application in crops, there was a negative balance of K in many soils. This was due to high removal of K from soils with continuous cropping. Results of the Long term fertiliser experiment of Ranchi with Soybean-Wheat sequence (Anonymous, 2013) reveal that because of high soil acidity (pH,5.3), partial factor productivity of Soybean or wheat crop was high with lime or FYM application. P application was a major factor for realization of good crop yields (Table 2). Further, in superimposed treatments with application of lime/FYM along with the farmers' practices i.e. application of urea alone or in combination with DAP or DAP+MOP, the results have clearly indicated that application of lime or FYM along with farmers' practice enhances the productivity by 6-728% as compared to their respective farmers' practice. Higher efficacy of these ameliorants was observed during *kharif* than *rabi* crops (Table 3). Lime application helped in raising soil pH, while organic manuring helped in chelating the acid forming cations and adding nutrients for better plant nutrition under N, NP and NPK treated plots.

**Table 2** : Partial factor productivity due to input use in Soybean-wheat cropping system in Acid soils of Ranchi

| Nutrient/ameliorants | Units | Soybean | Wheat |
|---|---|---|---|
| N | kg grain/kg N | -11.1 | -3.6 |
| P | kg grain/kg P | 21.4 | 79.2 |
| K | kg grain/kg K | 21.0 | 9.0 |
| Lime | kg grain/Q lime | 22.5 | 36.5 |
| FYM | kg grain/t FYM | 25.7 | 35.3 |

**Table 3** : Effect of lime/FYM in N,NP and NPK treated acid soils of Ranchi from 2002-03 to 2012-13

| Treatments | Soybean | | Wheat | |
|---|---|---|---|---|
| | Mean | % change | Mean | % change |
| N | 1.9 | | 2.7 | |
| N+Lime | 7.4 | 292 | 6.4 | 142 |
| N + FYM | 15.7 | 728 | 15.5 | 483 |
| NP | 5.6 | | 29.4 | |
| NP+Lime | 9.0 | 62 | 36.2 | 23 |
| NP + FYM | 16.6 | 198 | 38.8 | 32 |
| NPK | 19.3 | | 30.2 | |
| NPK+Lime | 21.4 | 10.6 | 35.0 | 16 |
| NPK+ FYM | 21.0 | 9.0 | 38.0 | 26 |

Good soil health implies that the system performs its ecological role in an unimpaired fashion. In recent years, 'soil quality' is being used in many cases in place of 'soil health'. This is to quantify soil health and give it a rigorous meaning by estimating the physical, chemical and biological properties and also assessment of larger scale features of the environment, such as bioclimate. Soil quality cannot be measured directly, but must be inferred from measuring the changes in its attributes of the ecosystem, referred to as indicators. Important soil quality indicators at micro and macro farm scale as suggested by Singer and Ewing (2000) are listed in Table 4. Each soil quality has an important role (Table 5) in influencing different soil processes and functions(Lal, 1998).

**Table 4 :** Important soil quality indicators at micro and macro farm scale

| Physical indicators | Chemical indicators | Biological indicators |
|---|---|---|
| Passage of air | BSP | Organic Carbon |
| Structural stability | Cation exchange capacity | Microbial biomass carbon |
| Bulk density | Contaminant availability | C and N/oxidiseable carbon |
| Clay mineralogy | Contaminant concentration | Total biomass |
| Colour | Contaminant mobility | Bacterial |
| Consistence (dry, moist, wet) | Contaminant present | Fungal |
| Depth of root limiting layer | Electrical conductivity | Potentially mineralizable N |
| Hydraulic conductivity | Exchangeable sodium percentage | Soil respiration |
| Oxygen diffusion rate | Nutrient cycling rates | Enzymes |
| Particle size distribution | pH | Dehydrogenase |
| Penetration resistance | Nutrient availability | Phosphatase |
| Pore conductivity | Nutrient content | Arlysulfatase |
| Pore size distribution | Sodium adsorption ratio | Biomass C/total organic carbon/ |
| Soil strength | | Respiration/biomass |
| Soil tilth | | Microbial community, fingerprinting |
| Soil structure | | Substrate utilization |
| Temperature | | Fatty acid analysis |
| Total porosity | | Nucleic acid analysis |
| Water holding capacity | | |

*Source* : Singer and Ewing (2000)

**Table 5 :** Major soil physical attributes/indicators, chemical and biological attributes and related processes.

| Attributes/indicators<br>**Physical attributes** | Processes and soil functions |
|---|---|
| **A. Mechanical** | |
| Texture | Crusting, gaseous diffusion, infiltration |
| Bulk density | Compaction, root growth, infiltration |
| Aggregation | Erosion, crusting, infiltration, gaseous diffusion |
| Pore size distribution | Water retention and transmission, root growth, and gaseous exchange |
| **B. Hydrological** | |
| Available water capacity | Drought stress, biomass production, soil organic matter content |

| | |
|---|---|
| Non-limiting water range | Drought, water imbalance, soil structure |
| Infiltration rate | Runoff, erosion, leaching |
| **C. Rooting zone** | |
| Effective rooting depth | Root growth, nutrient and water use efficiencies |
| Soil temperature | Heat flux, soil warming activity, and species diversity of soil fauna |
| Chemical Attributes | |
| pH | Acidification and soil reaction, nutrient availability |
| Base saturation | Adsorption and desorption, solubilisation |
| CEC | Ion exchange, leaching |
| Total/Available Nutrients | Soil fertility, nutrient reserves |
| Biological Attributes | |
| Soil organic matter | Structural formation, mineralization, biomass carbon, nutrient Retention |
| Earthworm population & Soil macrofauna activity | Nutrient cycling, organic matter decomposition, formation of soil Structure |
| Soil biomass carbon | Microbial transformations and respiration, formation of soil structure and organo-mineral complexes |
| Total soil organic carbon | Soil nutrient source and sink, biomass carbon, soil respiration and gaseous fluxes |

*Source*: Lal, 1998

One has to relate the soil quality with mineral composition of plants, which comprise our food. David Thomas (2003) reported from studies in U.K. that the fruit, vegetable, and cereal, which form the bulk of our diet, have become deficient in a range of minerals and trace elements compared to those 50 years ago (Table 6)

**Table 6 :** Changes in mineral content of different types of vegetables (27 varieties), Fruits (17 types) and Meat (10 cuts) measured between 1940 and 1991

| Mineral | Vegetable | Fruit | Meat |
|---|---|---|---|
| Na | -49% | -29% | -30% |
| K | -16% | -19% | -16% |
| P | +9% | +2% | -28% |
| Mg | -24% | -16% | -10% |
| Ca | -46% | -16% | -41% |
| Fe | -27% | -24% | -54% |
| Cu | -76% | -20% | -24 |

*Source* :David Thomas (2003)

These changes in mineral content is due to long term farming, imbalanced nutrient use, absence of micronutrients in the fertilization schedule, loss of soil organisms, changes in crop varieties etc. Ganeshmurthy (2009) argues that intrinsic soil fertility determines the carrying capacity of land and hence food supply and quality. Food shortages for a large percentage of the global population, malnutrition and mineral deficiencies need emphasis with a view to achieve food and nutritional security and a hunger-free world.

## Essential Plant Nutrients-role and Deficiency Symptoms

Soil is the reservoir of most of the essential plant nutrients. Plant mineral composition is dependent on the available nutrient status in soil and their supply to plants. Composition of nutrients in plants is especially relevant to meet the nutritional requirements of humanbeings and animals. Animals and human beings require Na, I, Cr, Co, F and Se besides the essential plant nutrients (Table 7)

**Table 7 :** Essential elements for plants, animals and human beings

| Species | Major Nutrients | Micronutrients |
|---|---|---|
| Plants | C, H, O, N, P, K, S, Ca, Mg | Fe, Mn, Zn, Cu, B, Mo, Cl, Ni. |
| Human beings | all the above + Na, Cl | Fe, Mn, Zn, Cu, Se, Cr, Mo, I, F |
| Animals | All the above + Na, Cl | Fe, Zn, Cu, Mn, Se, I, Mo, Co |

## Role of Essential Nutrients in Plant Nutrition

**Carbon:** The plant absorbs carbon dioxide directly from the atmosphere. This combines with water in the presence of light and forms the primary sugars, such as glucose and fructose (fruit sugar). Chlorophyll is the pigment, which absorbs radiant energy of the sun and brings about complex chemical synthesis of carbon dioxide and water resulting in simple sugars. This process is called photosynthesis or synthesis in light. Thus, by the combination of carbon with water, sun's energy is stored in the plant body, and the first carbohydrates are formed in the plant. From the carbohydrates, complex sugars, starches, hemicelluloses and celluloses are formed. These simple sugars also polymerise (chemically combine) into oils and fats. For instance, the soluble carbohydrates (sugars) decrease from 37.5 to 4.5 per cent during the ripening of sunflower seed. Same has been observed about oil in niger seeds.

**Oxygen:** Nutrient uptake by roots requires oxygen. It is a part of water as well as carbon dioxide. When water combines with carbon dioxide, oxygen is evolved:

$$6CO_2 + 6H_2O — C_6H_{12}O_6 + 6O_2$$

Thus, the oxygen evolved in the process equals the volume of carbon dioxide absorbed by the plants. This evolution of oxygen takes place in the process of photosynthesis. The process is reverse in respiration when a simple or a complex sugar, fat or oil breaks up. It requires oxygen and gives out $CO_2$.

$$C_6H_{12}O_6 + 6O_2 - 6CO_2 + 6H_2O$$

Six molecules of carbon in the glucose combine with six molecules of oxygen to form six molecules of carbon dioxide. In the process six molecules of water are formed. Thus, oxygen plays a dominant role in the processes of photosynthesis and respiration in plants.

**Hydrogen:** It is one of the most important elements in the nature. It readily combines with oxygen to form water and with carbon to form complex chemical organic compounds. The growth of plants would only take place if adequate quantity of water is supplied to meet the needs of hydrogen for synthesis of organic substances. Carbon, oxygen and hydrogen are present in essentially all organic compounds.When organic compounds either break up in the plant or decompose in the soil or atmosphere, the released hydrogen always combines with oxygen and forms water. Thus, the exchange of hydrogen takes place in either of the synthesis or decomposition (including respiration) processes.

**Nitrogen:** It is a constituent of proteins (enzymes), nucleic acids, porphyrins (chlorophyll, cytochromes) and alkaloids. Nitrate induces the formation of the enzyme 'nitrate reductase'. It is an essential part of carbohydrates, fats and oils besides proteins. Other constituents of proteins are oxygen, hydrogen and nitrogen, usually sulphur and sometimes phosphorus. The majority of the proteins have the following composition (in per cent):

| | |
|---|---|
| Carbon | 50.0-55.0 |
| Hydrogen | 6.5-7.3 |
| Nitrogen | 15.0-17.6 |
| Oxygen | 19.0-24.0 |
| Sulphur | 3.0-5.0 |

When proteins decompose through hydrolysis they give out amino acids; reversely, when proteins are formed or synthesized, the basic substances are amino acids.Nitrogen is the basic nutrient and makes up 1-4% of dry weight of plants and it forms chlorophyll, amino acids, proteins, alkaloids and protoplasm. In the plant sap, ammonia, nitrates and nitrites are found only in traces or in very small quantities. When the plant takes up large quantities of nitrogen from the soil, the colour of the plant changes to dark-green, indicating the increase of chlorophyll in the plant. With excess N, crop plants become more succulent and thus, susceptible to pests and diseases.

Source of Nitrogen to the plants are following:

i) Free living micro-organisms can fix N @ 16-50kg N/ha/year.

ii) Organic matter in the soil by decomposition produces 1-2 per cent N per ha and contribute 20-45 kg N/ha.

iii) Rainwater adds about 5-6 kg N/ha/year.

iv) Nitrogenous chemical fertilizers and organic manures / compost / vermi compost are important sources of N supply to crops.

**Phosphorus:** It is a constituent of the cell nucleus, essential for cell division and the development of meristematic tissues at the growing points. It makes 0.1 to 0.5% of dry weight of the plant.It is usually taken up by plants as orthophosphate ion ($H_2PO_4^-$), but can also be absorbed as secondary orthophosphates ($HPO_4^-$). Soil pH determines the ratio of uptake between the two. Phosphorus plays vital role in almost all plant processes that involve energy transfer. High energy phosphate,held as a part of the chemical structure of adenosine triphosphate (ATP),is the source of energy that drives the various chemical reactions within the plant. ATP is required for the process of photosynthesis. Light trapped by pigments (such as chlorophyll) is converted to chemical energy involving the high energy phosphate bonds. Carbohydrate metabolism,the process by which sugars and starch are broken down in growing plants,require P. Most of the movement of nutrients within the plant depends upon transport through cell membranes,requiring energy to oppose the forces of osmosis.The needed energy is provided by ATP. Organic P is about 20 to 80% of total P in soil. Most organic phosphates are present as ortho-phosphate esters.Phosphate can be split off from the esters by phosphatases originating from plants and/or microorganisms.Particularly,in the rhizosphere,the activity of phosphatases is high (Tarafdar and Jungk,1987) and hence the biological factor may contribute much to the soil phosphate potential (Tarafdar and Claassen,1988). Phosphate released from the organic phosphate pool by phosphatases may be directly absorbed by plant roots or exposed to P dynamics in soil.

**Potassium:** It is less able to absorb water and more subject to stress, when water is in short supply. Potassium does not become a part of the chemical structure of plants, it plays many important regulatory roles. Potassium is required to activate at least 60 different enzymes involved in plant growth. The accumulation of K in plant roots produces a gradient of osmotic pressure that draws water into the roots. Plants deficient in K have reduced rate of photosynthesis and the rate of ATP production and thus, slow down the processes dependent on ATP. If K is inadequate,sugar produced during photosynthesis is transported through the phloem to other parts of the plant at a slower rate due to less ATP production. Potassium plays a major role in the transport of water and nutrients throughout the plant in the xylem.When K supply is reduced, translocation of nitrates, phosphate, calcium, magnesium and amino acids is depressed. The enzymes responsible for synthesis of starch in leaves is activated by K. High levels of available K improves the quality, disease resistance, and feeding value of grain, horticultural and forage crops It constitutes 0.8 to 3.0% of dry matter in cereals.

**Sulphur :** Sulphur is a constituent of many proteins, and aids in the formation of chlorophyll and root growth. Plants having an abundant supply of sulphur develop dark-green leaves and extensive root system. In legumes, the nodular activity is appreciably increased by adequate supply of sulphur. S is a constituent of amino acids- cystine, cysteine,methionine and of coenzymes-thiamin, biotin and coenzyme A, Ferredoxins,compounds that functions in photosynthesis, contain S. Some special compounds and sometimes special odours contain S; mercaptans in radish roots and onions, disulfides in onion, polysulphides and sulphoxides in garlic and mustard oils.

Organic forms of Sulphur are dominant in soils. Only in arid soils, inorganic S is more than organic S. Organic S occurs as reduced S, in which the S is bound to a C atom (C-S), and in sulphate esters. Unlike nitrate, sulphate is adsorbed to soil colloids in substantial amounts and therefore adsorbed sulphate may contribute to S potential of soils. Sulphate adsorption is strong at low soil pH and decreases as pH increases. For this reason, liming may increase the potential of plant available sulphate.

**Calcium:**Calcium is essential for the formation of cell-walls, as calcium pectate forms part of the middle layer of the cell-wall.Ca is a constituent of phytin and pectin (important for mechanical strength). Ca is present in enzyme amylase and precipitates, such as calcium carbonate,calcium oxalates. Ca activates some enzymes (alpha-amylase, phospholipase, AT Pases) and can act as a competitive inhibitor for Mg. (Schimansky, 1981). In root-tips, calcium is very essential for the meristematic activity or formation of new tissues. It also helps to keep up sustained activity of the nodule bacteria in legumes. $Ca^{2+}$ demand of crops differ among plant species, generally dicots have higher $Ca^{2+}$ demand than monocots. This is because the latter binds comparatively low amounts of $Ca^{2+}$ in the cell wall.

Besides its direct nutrient value, calcium when applied to acid soils increases the availability of other nutrients, like phosphorus, nitrogen and molybdenum. Excess of calcium in calcareous soils, depresses the uptake of potassium and magnesium. These are secondary effects of calcium on plant growth. Acidic organic soils (Histosols) and highly weathered soils of humid regions (Oxisols) commonly have a low $Ca^{2+}$potential that is based on the $Ca^{2+}$ demand of plants or $Ca^{2+}$ demand of soils.Soil structure depends to a large extent on the degree of $Ca^{2+}$ saturation. About 70 to 80% of the CEC should be occupied by $Ca^{2+}$ in order to maintain a satisfactory soil structure.

**Magnesium:**The most important $Mg^{2+}$ bearing minerals are olivine, serpentine, biotite, pyroxene, amphibole, and dolomite. $Mg^{2+}$ release from mica is specially high in acid soils. Released $Mg^{2+}$ may be quickly adsorbed to soil colloids and this fraction represents the exchangeable $Mg^{2+}$. It is often used as an indicator for available soil $Mg^{2+}$. $Mg^{2+}$ can be leached from the coarse textured soils. Entisols (young soils) are generally rich in $Mg^{2+}$. Highly weathered soils under humid tropics and also the temperate zone often have low exchangeable $Mg^{2+}$. Mg is a constituent of chlorophyll, phytin,and pectin. Mg is an activator of many enzymes, especially phosphorylating enzymes (connected with energy metabolism), bridging enzymes (synthesis of RNA). It is antagonistic to K absorption.

**Iron:** Iron is present in large quantity in soils, but its solubility is very low /and plants would suffer from its deficiency, if other factors did not come into play. These are the production of numerous siderophores by microbes (Crowley et al., 1991) and higher plants. Siderophores are organic compounds, which dissolve $Fe^{3+}$ from soil minerals and which form $Fe^{3+}$ complexes soluble over a broad pH range (Chesworth Ward, 2008). Fe is a constituent of many compounds; iron porphyrins, such as cytochromes, cytochrome oxidases, catalases, peroxidases. Iron is an activator of enzymes such as aconitase. Iron deficiency in crops is common in calcareous soils (lime induced chlorosis). Fe toxicity is often observed in submerged soils due to reduction and enhanced solubility of iron under reduced conditions.

**Manganese:** It is an activator of many enzymes in energy metabolism (phosphorylation), arginase in photosynthesis, formation of chlorophyll. It has role in synthesis of proteins, carbohydrates, lipids and synthesis of Vitamin C. Mn is present in soils as Mn oxides and Mn bearing minerals. Its availability to plants depends on soil pH, redox potential, and microbiological activities in soils. Decrease in redox potential in flooded soils increase soluble Mn. Under aerobic conditions, Mn solubility decreases with increase in soil pH by a factor of 10 for each unit of pH. This insufficient Mn supply to crops is likely to occur in alkaline soils. High pH associated with high soil organic matter can cause Mn deficiency.

**Boron:** One of the most marked effects of boron deficiency observed is the restricted development of nodules on the roots of legumes. Very little nitrogen is fixed in these nodules. Besides, its deficiency influences the growing points of stems, buds and roots. The tissues carrying the minerals and water from the soil to the leaves are also disorganized in the absence of boron. The leaves become brittle. Boron is known to stimulate germination of pollen and tube growth.

Young soils and marine sediments are generally rich in B.Boron may also accumulate in alkaline soils even to levels, which is toxic to plants. Highly weathered soils in humid areas are often low in B. On such soils, B is leached out of the root zone.

**Zinc:** In a general way, zinc is associated with the development of chlorophyll in leaves and a high content of zinc is correlated with a high amount of chlorophyll. Zn is a constituent of some dehydrogenases (alcohol dehydrogenases, lactic dehydrogenases). Zn also functions in the formation of auxins, in carbonic anhydrase and peptidases. Zn concentration is low in calcareous soils. Zn solubility decreases with rise in pH. A substantial amount of Zn is complexed by organic molecules and hence soluble in the soil solution. Zn deficiency is common in waterlogged rice soils. In arid areas and low temperature conditions, Zn deficiencies of crops are common.

**Copper:** Cu is a constituent of enzymes (oxidases, functions through change in valence) such as lactases, phenolases,ascorbic acid oxidases,and cytochrome oxidases, superoxide dismutase, amine oxidases. It is also a constituent of plastocyanin, a compound that functions in photosynthetic electron transport. Copper in soils is exclusively present in bivalent form. A substantial proportion of Cu in soils is present in organic complexes. Copper deficiency in crops occur in humic podzolic soils.

**Molybdenum:** Mo is a constituent of enzymes involved in nitrogen fixation (nitrogenase) and nitrate reduction (nitrate reductase). The presence of molybdenum is very essential for the fixation of atmospheric nitrogen the roots of legumes by nodule bacteria. Mo is present in soils as $MoO_4$ (oxyanionic form). Soluble molybdate ions, $HMO_4^-$ and $MO_4^{2-}$ can be strongly adsorbed to Fe/Al oxides/hydroxides. The process is a ligand exchange analogous to that of phosphate. The adsorption is stronger, lower the soil pH. For this reason Mo deficiency mainly occurs on acid soils.

**Nickel:** Nickel is chemically related to iron and cobalt. Its preferred oxidation state in biological systems is $Ni^{2+}$, but it can also exist in the states of $Ni^+$ and $Ni^{3+}$, it forms stable complexes or ligands with cysteine, citrate and enzymes. Urease is the enzyme containing Ni. It plays an important role in nitrogen metabolism and application of Ni during foliar spray of urea, prevents urea toxicity in plants and improve crop yield. In most plants, Ni content in the vegetative organs is in the range of 1-10μg $g^{-1}$ dry weight. Nickel is readily mobile in the xylem and phloem and in some plant species preferentially translocated into the seeds. So far, there is no clear evidence of Ni deficiency in soils.

## Deficiency Symptoms of Nutrients in Plants

Deficiency symptoms can be categorized into five types.

i) Chlorosis, which is yellowing, either uniform or interveinal of plant leaf tissue due to reduction in the chlorophyll formation.

ii) Necrosis, or death of plant tissue.

iii) Lack of new growth or terminal growth resulting in rosetting.

iv) An accumulation of anthocyanin and / or appearance of a reddish colour.

v) Stunting or reduced growth with either normal or dark green colour or yellowing.

**Nitrogen:** The nitrogen-deficient plants show a general yellowing of old leaves, blades of cereals tend to become stiff and erect. The lower leaves turn yellow and in some crops they quickly start drying up as if suffering from shortage of water.

**Phosphorus:** Purplish discoloration or darker leaves due to anthocyanin formation. Stiffness of leaf blades with tips bent outwards. In case of P deficiency, symptoms are first seen in older leaves.

**Potassium:** Deficiency symptoms occur first in older leaves. Yellowing followed by necrosis, at tip and margin of leaves (marginal scorching), wilting of foliage, tendency to lodge in cereals. In some vegetables, white speckling of leaf blades.

**Sulphur:** Light green to yellow colour of young leaves (slow mobility), sometimes affecting the whole plant. Plants become spindly and small.

**Magnesium:** Characteristic interveinal chlorosis,later necrosis of mature leaves, upward curling of leaves along margins.

**Calcium:** Death of growing points and root tips (immobile element), buckling of stems, dark foliage, shedding of blossoms and buds,water soaked, discolored areas on fruits (blossom-end rot of tomatoes, peppers, melons; bitter pit or cork spot of apples ). Finck(1992) has provided a key to nutrient deficiency symptoms in plants (Table 8)

**Table 8 :** A brief key for nutrient deficiency symptoms

| Deficiency symptoms | Nutrient |
|---|---|
| Symptoms appearing first on older leaves | |
| Chlorosis starting from leaf tips | N |
| Necrosis on leaf margins | K |
| Cholorosis between veins | Mg |
| Brownish, greyish, whitish spots (e.g.cereals) | Mn |
| Reddish colour on green leaves or stem | P |
| Symptoms appearing first on younger leaves | |
| Mottled yellow-green leaves with yellowish veins | S |
| Mottled yellow-green leaves with green veins | Fe |
| Brownish black spots (e.g. legumes,potato) | Mn |
| Youngest leaf has white tip | Cu |
| Youngest leaf is brownish or dead (e.g.beet) | B |

**Micronutrients:** The deficiency symptoms of micronutrients vary widely with crop species. Without going into the details of such variations, a brief account is presented below (Chesworth Ward,2008).

**Zinc:** Mottled leaves, interveinal chlorosis, decrease in stem length causing rosetting of terminal leaves, dieback of twigs, small leaves ('little leaf' disease)

**Boron:** Stunted growth, chlorosis, bronzing, curling, and wilting of new leaves, death of terminal bud, cracking of stems and roots.

**Manganese:** Interveinal chlorosis in youngest leaves, with gradual colour change towards green veins (in contrast to Fe deficiency)

**Iron:** Interveinal chlorosis of young leaves, veins remain green. In severe cases, the entire leaf turns chlorotic and changes from yellow to whitish. Leaves of some plants turn necrotic. Injury to terminal bud causes development of shoots from dormant side buds.

**Copper:** Wilting, death of leaf tips, dieback of terminal shoots in trees, poor pigmentation, gum formation.

**Molybdenum:** Cupping or rolling of leaves, reduced development of leaf blades (whip tail) in some plants (cauliflower), some symptoms similar to N deficiency.

Special names have been given in crops for micronutrient deficiencies. These are:

Fe : lime induced chlorosis

Mn : gray speck of oats

B : heart rot of sugarbeets, cracked stem of celery, white top of alfalfa, brown rot of cauliflower.

Zn : little leaf and rosette disease of fruit trees, khaira of rice.

Cu : exanthema in pear (gums exuded from barks)

Mo : whiptail of cauliflower, yellow spot of citrus.

Ni : mouse ear of pecans

Prominent nutrient deficiency symptoms in plants are summarized below:

| **Nutrient** | **Colour change in lower leaves** |
|---|---|
| N | Plants light green, older leaves yellow |
| P | Plants dark green with purple cast, leaves and plants small |
| K | Yellowing and scorching along the margin of older leaves |
| Mg | Older leaves have yellow discolouration between veins-finally reddish purple from edge inward |
| Zn | Pronounced interveinal chlorosis and bronzing of leaves |
| **Nutrient** | **Colour change in upper leaves** (Terminal bud dies) |
| Ca | Delay in emergence of primary leaves, terminal buds deteriorate |
| B | Leaves near growing point turn yellow, growth buds appear as white or light brown, with dead tissue. |
| **Nutrient** | **Colour change in upper leaves** (Terminal bud remains alive) |
| S | Leaves including veins turn pale green to yellow, first appearance in young leaves. |
| Fe | Leaves yellow to almost white, interveinal chlorosis at leaf tip |
| Mn | Leaves yellowish-gray or reddish, gray with green veins |
| Cu | Young leaves uniformly pale yellow. May wilt or wither without chlorosis |
| Mo | Wilting of upper leaves, then chlorosis |

## Some Photographs on Nutritional Deficiencies of Crops

Tomato plants with N on left side and without N on right side

Leaves of tomato showing N deficiency

Very severe N deficiency in older leaflets of tomato plant, right leaflet fertilized with nitrogen.

Phosphorus deficiency in Crucifers

Leaves showing potassium deficiency in cauliflower before curd development

Leaves and leaflets of tomato plants with marginal chlorosis & differentially developed symptoms of K deficiency

Leaves of potato showing S deficiency

Leaves of Sugar beet showing S deficiency

a. Tomato plants with marginal chlorosis & necrosis and forward curled leaf margins due Ca deficiency.

b. Sprout top of a tomato plant with softened and bent down stem due to Ca deficiency.

c. Tomato leaf with vein browning due to Ca deficiency.

Tomato plant with 'blossom end rot' in fruits due to Ca deficiency induced by high P&K in an acid soil.

Boron deficiency in Cauliflower

Boron deficiency in fruits of Tomato

Boron deficiency in Papaya

Boron deficiency in Litchi

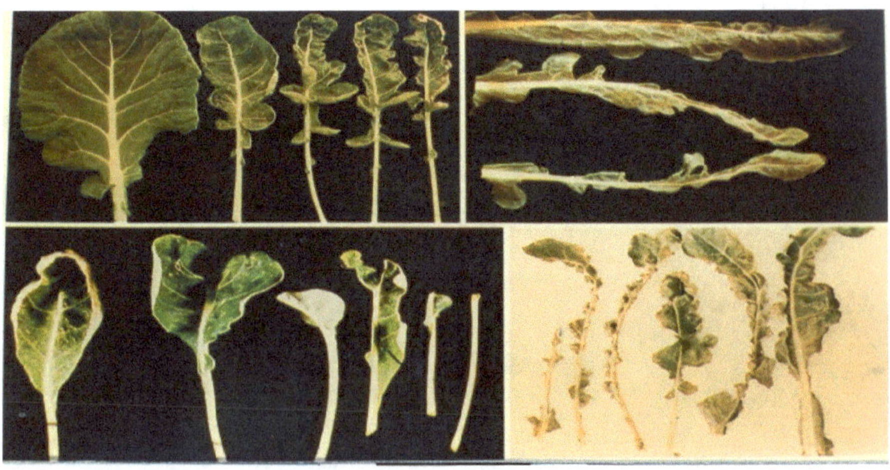

Distorted and reduced leaf area, whiptail formation, leaf cupped, tips and margins developed partly dead and necrotic patches showing characteristic symptoms of Molybdenum deficiency in cauliflower. Top left corner-leaf of a healthy plant.

Pecan leaves exhibiting typical rounded, blunt leaflet tips.

## Soil Nutrients vis-a-vis Human/Animal Health

Human beings need most of the nutrients available in soil. Nutrients available in soil, thus, must be translocated to plants or food materials used for consumption. Mining of nutrients from soils, nutrient imbalances, soil related constraints and non-application of plant nutrients have made the problem worse causing health problems in animals and human beings. This is the reason that soil health needs to be linked to food quality, food safety, balanced nutrition and health.

Human beings and animals need nitrogen compounds (protein, amino acids), which comes from intake of cereals, pulses, vegetables, fruits etc. Animals especially dairy cows require Phosphorus, since milk contains high P. Adult human beings have 0.9 kg P in the body (bones and teeth primarily). Potassium helps in higher Vitamin C level in crops. Calcium helps in bone, teeth and cellwall formation in human beings and animals. Mg deficient diet in animals results in tetany in extreme cases, muscle spasm etc. Sulphur is essential for digestion, purification of the system, S-containing aminoacids and bile secretion.

Among the micronutrients, Zn and Fe deficiency diseases in human beings and animals are common and increasingly observed. More than 25% of the population of India is under the risk of low Zn and Zn-related diseases. About 50% of Indian soils are deficient in available Zn. Dietary intake of less than 10mgZn/day in human beings is regarded as deficient. Zn deficiency has adverse effect on the immune system and increases the incidence of diarrhoea and pneumonia. Zn deficiency in children is common and is associated with stunting. Iron deficiency is common in upland soils of dry regions. Iron is important for muscle, brain, and red blood cells. Iron deficiency anaemia with low haemoglobin is common in women and children.

Singh (2009) citing a study by Narwal *et al.* (2006) indicated that in Haryana, most of the districts have low Zinc and Iron in soils, while Mn deficiency was emerging fast in intensively cultivated alkaline soils. When feed and fodders produced on such deficient soils were fed to buffaloes, they showed higher percentage of deficiencyin their blood serum, hair and milk (Table 9). The deficiencies of Ca,P,Cu and Zn were found to be widespread in milk and hair. Deficiency of minerals is common. For example, Ca deficiency ranged from 23 to 57%, that of P, 24 to 69% and Cu, 33 to 64% in blood serum,hair,and milk samples.

**Table 9 :** Percentage of mineral nutrient deficiency in blood serum,hair,and milk in cattle in Haryana.

| District | % deficiency in blood serum | | | % deficiency in hair | | | % deficiency in milk | | | |
|---|---|---|---|---|---|---|---|---|---|---|
| | Ca | P | Cu | Ca | Zn | Mn | Ca | P | Zn | Cu |
| Gurgaon | 80 | 19 | 42 | 27 | 44 | — | 30 | 28 | 51 | 48 |
| Rewari | 61 | 48 | 51 | 40 | 53 | 30 | 45 | 65 | 33 | 35 |
| Faridabad | 73 | 06 | 07 | 41 | 34 | — | 35 | 24 | 64 | 71 |
| Bhiwani | 35 | — | 37 | 27 | 96 | 99 | 66 | 62 | 90 | 29 |
| Rohtak | 29 | — | 46 | 55 | 81 | 67 | 71 | 80 | 63 | 25 |
| Hisar | 37 | 13 | 34 | 36 | 31 | 30 | 24 | 51 | 30 | 06 |

## Strategies to Check Plant Nutrient Disorders

Livelihood of a vast percentage of our population in India depends on agriculture. Agriculture is also the major supplier of nutrients for human beings and animals through plants. Thus, it appears that by improving the nutrient content in plants, or by improving the availability of nutrients in soil and improving their uptake by growing plants, one can solve the problem to a considerable extent. Some of the measures in this direction are listed below:

- Improved cropping systems in different agro-ecological zones & their adoption by farmers.
- Improved nutrient content in edible parts of the plant through screening & genetic manipulation in crops.
- Integrated plant nutrient management in crops and cropping systems.
- Improving the soil organic matter content.
- Amelioration and Management of problem soils.
- Micronutrient fortification and supplementation.
- Dietary modification or change in food habit.
- Motivating farmers to produce nutritious and diverse agricultural products.
- Creating awareness among consumers about consequences of malnourished seeds and poor diets on health.

## References


Anonymous (2013) Annual progress report AICRP-LTFE of Ranchi centre, BAU., pp1-39.

Arnon, D.I. and Stout, P.R. (1939) An essentiality of certain elements in minute quantity for plants with special reference to copper. Plant Physiology, 14, 371-375.

Asher, C.J. (1991) Beneficial elements, functional nutrients and possible new essential elements. In Mortvedt, J.J., Cox, F.R., Shuman, L.M., and Welch, P.M.eds. Micronutrients in Agriculture, 2nd edition, Madison, WI: Soil Science Society of America, pp703-723.

Chesworth Ward(2008) Encyclopedia of Soil Science (ed.). Springer publication, Netherlands.

Crowley, D.E., Wang, Y.C., Reid,C.P.P. and Szaniszlo, P.J. (1991) Mechanisms of Iron acquisition from siderophores by microorganisms and plants. In Chen, Y. and Hadar, Y. (eds.) Iron Nutrition and interactions in Plants. Dordrecht : Kluwer Academic Publishers. pp213-233.

David Thomas (2003) A study of the mineral depletion of the foods available to us as a nation over the period 1940-1991. Nutrition and Health 17,85-115.

Finck,A.(1992) Fertilisers and their efficient use. In IFA World Fertiliser Manual. Paris, France. 1-36.

Ganeshmurthy, A.N.(2009) Linking human health to soil health. Bull. Indian Soc. Soil Sci.27, 15-40.

Lal, R (1998) Basic concepts and global issues:Soil quality and agricultural sustainability. in Soil quality and Agricultural sustainability (R.Lal ed.) Ann. Arbor. Science, Chelssea, MI, USA, pp 3-12.

Nambiar, K.K.M. and Abrol,I.P.(1989) Long term fertiliser experiments in india-an overview. Fert. News 34, 24-38.

Narwal, R.P., Ramkala and Malik, R.S. (2006) Delineation of micronutrient deficient area. Annual progress report of AICRP-micronutrients, CCSHAU, Hisar Centre, pp1-85.

Singer, M.J. and Ewing, S (2000) Soil Quality. In : Handbook of Soil Science (Summer ed.) CRC Press, Boca Raton, FL, pp 271-298.

Singh, M.V. (2009) Effect of trace element deficiencies in soils on human and animal health. Bull. Indian Soc. Soil Sci. 27, 75-101.

Tarafdar, J.C. and Claassen, N. (1988) Organic Phosphorus compounds as a source of P for higher plants through phosphatases produced by plant roots and microorganisms. Biol. Fert. Soils, 5, 308-312.

Tarafdar, J.C. and Jungk, A. (1987) Phosphate activity in the rhizosphere and its relation to the depletion of soil organic Phosphorus. Biol. Fert. Soils 3, 199-204.


❑❑❑

Chapter - 3

# Soils of India-Nutrient Status and Management Concerns

**A K Sarkar**

Soils of India vary widely among different regions. The first soil map of India was prepared and classified on the basis of existing knowledge of Russian pedological principles. Govind Rajan (1971) and Govind Rajan and Rao Gopal (1978) have discussed in detail the later developments on the preparation of soil map of India and giving the equivalents of different soil groups in US Comprehensive System of Soil Classification. Soil resource mapping was initiated in 1986 and in a span of 10 years (1986-1996), a detailed database in terms of soils, their area and extent, characteristics and grouping following soil taxonomy (Soil Survey Staff, 1999) was created. The present chapter on "soils of India-nutrient status and management concerns" takes into account the following aspects:

1. Major soil types of India
2. Soil degradation
3. Nutrient mining
4. Soil fertility and nutrient availability trends
5. Carbon sequestration
6. Nutrient Management strategies

## 1. Major Soil Types of India

Sehgal *et al.* (1987) at National Bureau of Soil Survey and Land Use Planning, a premier soil research institute of the ICAR, through their pioneering studies for several years provided an insight on the soils of the country and their taxonomy. Major soil types of India can be grouped as: Alluvial, Black, Red, Laterite and lateritic soils. Besides, we have large areas under saline and alkali soils, acid soils, desert soils, mixed red and black soils, forest and hill soils etc. (Fig.1 to 5). The area, occurrence, distribution, characteristics and constraints with major soil types have been given in Table 1.

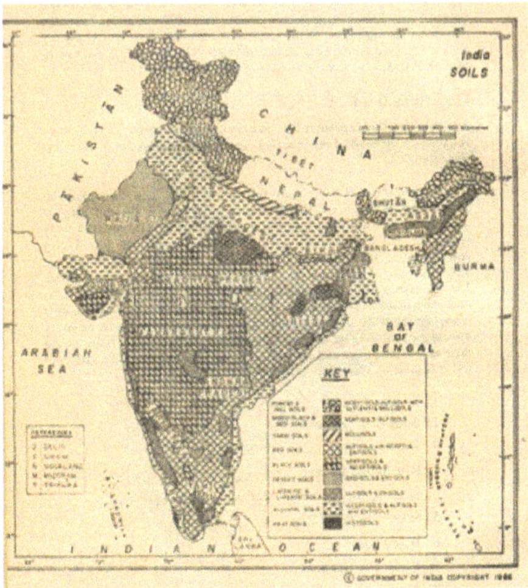


**Fig. 1 :** Major Soils of India

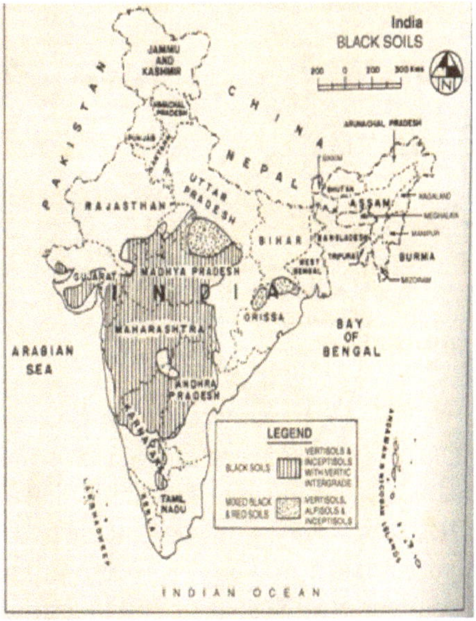


**Fig. 2 :** Black soils of India

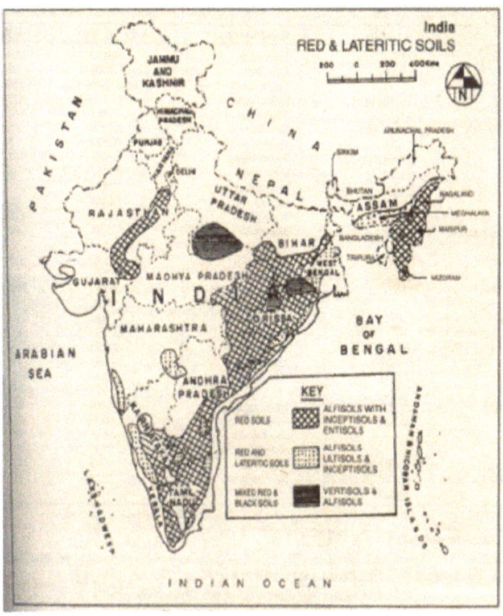


**Fig. 3 :** Red and laterilic soils of India

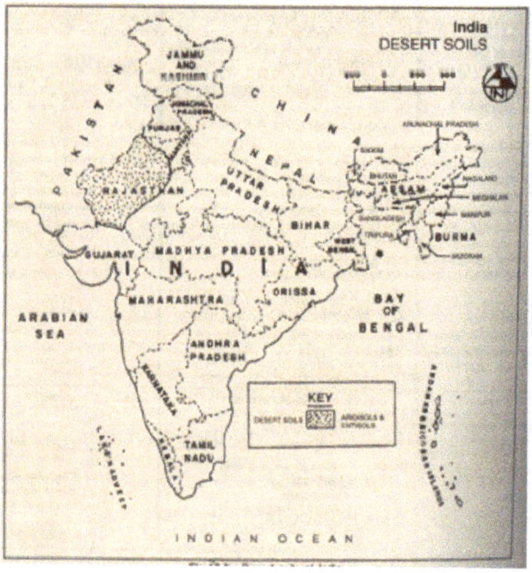


Fig. 4 : Desert-soils of India

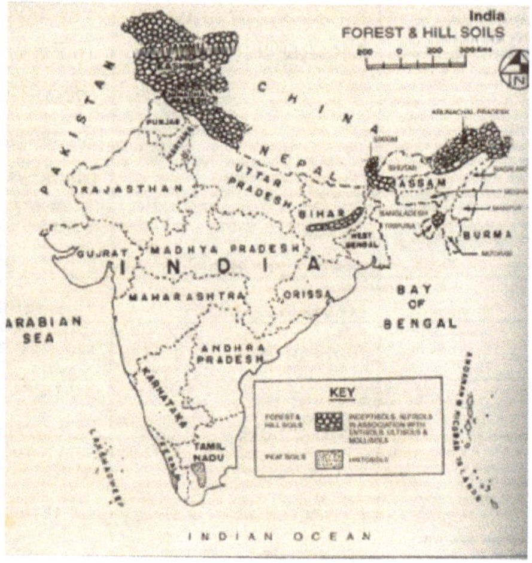


Fig. 5 : Forest and Hill soils of India

**Table 1 : Major soil types of India and their salient features**

| | **Alluvial soils** | **Black Soils** | **Red soils** | **Laterite and lateritic soils** |
|---|---|---|---|---|
| **Area (Mha)** | **75.0** | **72.0** | **45.0** | **25.0** |
| Distribution | Northern, north-western and north eastern part of India, including Punjab, Haryana, Uttar Pradesh, Delhi, Bihar, Uttarakhand, West Bengal, Assam and coastal areas of the country | Dominant in central, western and southern states of India | Predominant in southern part of peninsula in the states of Andhra Pradesh, Tamilnadu, Karnataka, Maharashtra, In eastern and north-eastern states of Odisha, Jharkhand, West Bengal, Assam. | Maharashtra, Andhra Pradesh, Tamilnadu, Karnataka, Odisha, Jharkhand and north-east region |
| Parent material | Alluvium | Volcanic rocks such as Basalt and other meta-morphic rocks rich in Smectites | Granite, Gneiss | Basic rocks . Soils formed are rich in sesquioxides, devoid of bases and primary silicate minerals |

*Contd...*

| Orders | Entisols, Inceptisols, Alfisols and Aridisols | Vertisols (deep black soils). Inceptisols and Entisols (shallow black soils) | Alfisols, Inceptisols, Ultisols Entisols | Oxisols, Ultisols Inceptisols |
|---|---|---|---|---|
| Characteristic | These soils are generally variable in colour, texture, moisture regime and organic matter. Soil profile development varies with age (recent, young, old etc.) from least developed A-C to well developed A-Bt-C horizons. | The soils are highly argillaceous with clay content varying from 30 to 80 per cent. These soils have high CEC, pH ranging from 7.8 to 8.7. Dark in colour due to clay-humus complexes and due to titaniferrous magnetite minerals. High water and nutrient retention capacity. Soils and are extremely hard when dry and sticky and plastic when wet | Red to yellow in colour. Red colour is due to ferric oxide (haematite),while yellow colour is due to limonite (hydrated iron oxides) Highly variable in texture, shallow (uplands) to deep (valleys), well drained, low CEC. | Soils are leached, poor in bases (Ca,Mg, Na,k) and silica. Acidic with accumulation of sesquioxides Low in CEC, Kaolinite is the dominant clay mineral with base saturation <40%. Lateritic soils in dry areas have high base saturation (> 40%) |
| Soil fertility | Rich in K, sufficient in P, poor in N and organic matter content. These are generally alkaline in reaction, but may be acidic in high rainfall regions. Alluvial soils are best agricultural soils and are used for growing most agricultural crops. | These soils are inherently very fertile and predominantly used for growing cotton, millet, sorghum, soybean and other crops. | Soils are deficient in N, P, K, organic matter and are poor in base saturation (rich in Kaolinites). Soils are poor in water and nutrient retention capacity. Poor in fertility. | Soils are deficient in P, K, Ca, Zn, B with low pH. |
| Nutrient management | These need special management | These soils are difficult to manage | Soils are suited for most crops, | Need proper management |

*Contd...*

| issues | practices to overcome second generation problems such as secondary & micronutrient deficiencies, salinization. | because of poor infiltration rate, poor drainage and calcareous nature. Fixation of P and Zn is a problem. | horticulture and plantation crops with good management | of nutrients for good crop yields. |
|---|---|---|---|---|

Bhattacharya et al. (2009)

Besides these major soil groups, we have a wide variety of soils. About 7Mha of land surface is covered by salt affected soils, which produce low and uneconomical yields of crops (Fig. 6). Agro-ecological sub regions in the coastal areas of India have 10.8Mha area under coastal soils. Such soils are characterised by excess accumulation of soluble salts and alkalinity in soil, pre-dominance of acid sulphate soils, periodic inundation of soil surface by tidal water and eutrophication and hypoxia. In India, these soils occur in coastal plain of Kerala and deltaic area of Sunderbans of West Bengal. Sahoo and Sarkar (2013) have presented a detailed account of Acid soils of India covering over 25Mha of land surface with pH below 5.5 (Fig. 7). Forest and hill soils occupy about 75Mha mainly in the states of Himachal Pradesh, Jammu & Kashmir and Uttarakhand (Fig. 5). Desert soils occupy an area of about 29 Mha in the north-west India (Rajasthan, Gujarat, Punjab). These soils have a thick mantle of sand in the surface deposited by aeolian action (Fig. 4). This hot arid region with a growing period of <60 days have poor profile development. Soil resources of the country and their management constraints are presented in Table 2.

**Table 2 :** Soil resources of India and their constraints

| Soil order | Land area (Mha) | % of total area | soil related constraints |
|---|---|---|---|
| Inceptisol | 130.37 | 39.8 | Soil erosion, nutrient imbalance, low soil organic matter. |
| Entisols | 92.13 | 28.0 | Soil erosion, nutrient imbalance, low soil organic matter. |
| Vertisols | 27.96 | 8.5 | massive structure, poor tilth, drought Stress, water erosion |
| Alfisols | 44.29 | 13.5 | weak soil structure, crusting, compaction, water erosion |
| Aridisols | 14.07 | 4.3 | Drought stress, nutrient depletion, wind erosion, desertification, salinization. |
| Ultisols | 8.25 | 2.5 | Water erosion, nutrient imbalance, acidification, P fixation. |
| Mollisols | 1.32 | 0.4 | high soil organic matter |
| Histosols | 0.002 | — | high soil organic matter |
| Others | 9.67 | 2.95 | |
| Total | 328.06 | 100 | |

Velayutham and Bhattacharya (2000)

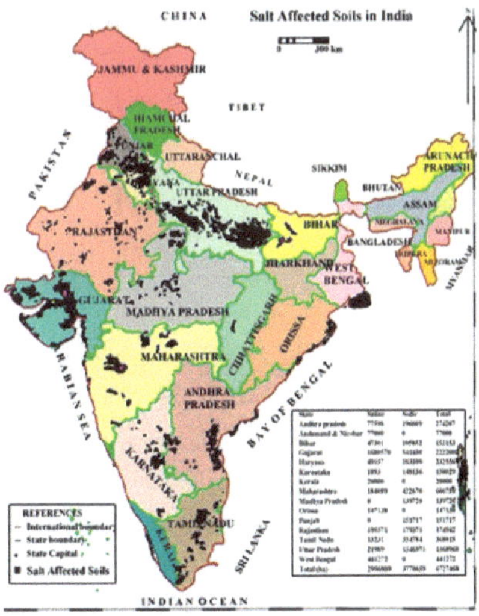

Fig. 6 : Salt-affected soils of India

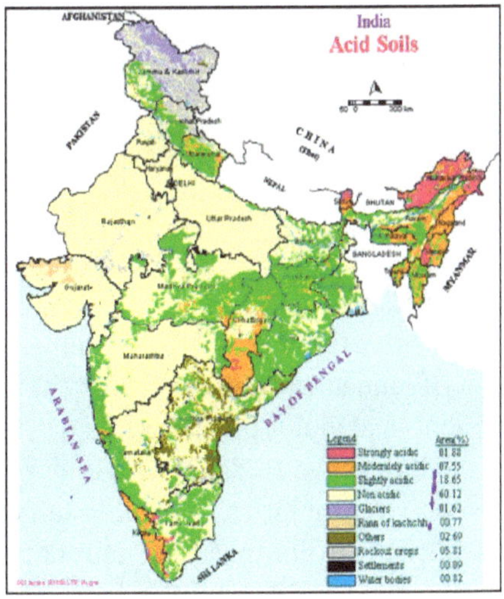

Fig. 7 : Acid soils of India

## 2. Soil Degradation

Soil degradation refers to decline in soil productivity through adverse changes in nutrient status and soil organic matter, structural attributes and concentration of electrolytes and toxic chemicals. It refers to decrease in capacity of soil to produce crops with good quality as a result of one or more degradative processes. Soil degradation can have physical, chemical or biological impact (Table 3). Physical degradation is due to erosion of top soil by water or air and loss of nutrients. Chemical degradation of soils involve a number of processes, such as loss of nutrients due to faulty management practices, loss of soil organic matter due to continuous cropping and/or non-addition of organic manures in soil, toxicity of nutrients, such as Iron and Aluminium in highly acidic soils, high sodium in alkali soils and excess of soluble salts in saline soils etc. There has been a steady rise in the use of chemical fertilisers of Indian soils to augment crop production. This is due to introduction of high yielding /hybrid crop varieties mainly in the 40% of the irrigated area of the country. Also, with increase in area after availability of assured irrigation facilities, the cropping intensity is rising steadily. Food demand is also increasing with rise of population. But, nutrient use in many areas continue to be imbalanced, with greater use of N, less use of K, S and micronutrients, such as Zn and B. Availability of organic manures and its regular application by farmers is showing a decline. This has resulted in a decline in soil organic carbon, soil microflora and microfauna. Inherent soil fertility thus, has shown a decline over years in major crop production zones of the country.

**Table 3 :** Soil degradation and its impact

| Soil Degradation | | |
|---|---|---|
| **Physical** | **Chemical** | **Biological** |
| compaction, desertification, erosion (water & wind) | fertility, pollution/toxicity imbalanced nutrient use, acid, saline, sodic soils | reduction in soil organic matter, reduction in biodiversity |

As per the latest estimates of NBSSLUP (2008), about 120.8 Mha is the total degraded area in the country, which is 36.5% of the total geographical area. These include eroded soils, waterlogged soils, area affected by salinity, alkalinity/acidity (Fig. 8). Sharma and Singh (2009) outlined the problems of degraded soils in different regions of the country (Table 4). More than 33% of the total geographical area is affected by soil erosion with soil loss of more than the permissible rate of 10t/ha/year. The states of Arunachal Pradesh, Assam, Chhattisgarh, Jharkhand, Madhya Pradesh, Manipur, Meghalaya, Nagaland, Uttar Pradesh, Uttarakhand have more than 50% area affected by soil erosion (Fig. 9) The area under severe soil erosion (>40t/ha/year) is about 11%. In quantitative terms, about 5.3 billion tonnes of soil is lost in India at an average rate of 16.3t/ha/year. About 8Mt plant nutrients are washed away along with eroded sediments.

**Table 4 :** Area under degraded soils in India (Mha)

| Sl No. | Type of degradation | Area (Mha) |
|---|---|---|
| 1. | water erosion(> 10 t/ha/yr) | 73.27 |
| 2. | wind erosion (aeolian) | 12.40 |
| 3. | salt affected soils | 05.44 |
| 4. | salt affected & water-eroded soils | 01.20 |
| 5. | acid soils (pH < 5.5) | 05.09 |
| 6. | acidic (pH,5.5) & water-eroded soils | 05.72 |
| 7. | degradation due to mining & industry | 0.19 |
| 8. | waterlogged & marshy lands | 0.97 |
| 9. | area under open forest (<40% canopy) | 16.53 |
| | **Total** | **120.81** |

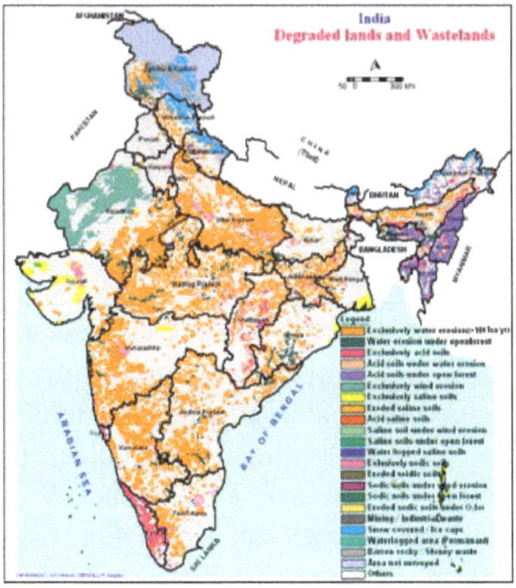


**Fig. 8 :** Degnaded and waste lands of India

**Fig. 9 :** Magnitude of soil erosion in India
*Source* : NBSS&LUP (2008)

The effect of soil degradation on crop productivity has been analysed by various workers (Sharda *et al.*, 2009). The analysis reveal that India suffers a loss of about 13.45Mt for the major rainfed cereal, oilseed and pulse crops due to erosion by water with maximum loss under cereals (66%) followed by oilseeds (21%) and pulses (13%). In monetary terms, the loss works out to be about Rs.112 billion.

About 12Mha area is flood prone and waterlogged in Vertisols of central India and Inceptisols and Entisols of eastern India , where crop productivity is severely affected. Salt affected soils are distributed extensively in Uttar Pradesh, Gujarat, Maharashtra, West Bengal, Tamil Nadu and Andhra Pradesh. Salinity problems are increasing in canal command areas, resulting in adverse impact on crop yields. On the other extreme, about 25 Mha of acidic soils ($pH<5.5$) located mainly in the eastern and north-eastern region suffer from $Al^{3+}$ toxicity, nutrient imbalance and resultant low productivity. The chemistry,distribution and management of such soils have been described in detail recently by Sarkar (2013). To sum up, over-exploitation and unscientific management of limited soil resource without caring for long term sustainability, has resulted in different kinds of land degradation. Poverty in India is both a cause and a consequence of increasing soil degradation and resultant decline in productivity.

### 3. Nutrient Mining

Satish Chander (2010) in his remarks under 'frank notes' has pointed out the emerging soil health issues of Indian soils. He says that Indian soils are under threat of different degradation processes, particularly nutrient mining, which has serious implications in terms of deterioration of soil health, which weakens the foundations of sustainable agriculture. Against an estimated annual removal of 34 Mt of nutrients(NPK) from soil, the replenishment through fertilisers is only 26Mt, leaving a net annual deficit of 8Mt of nutrients, which keeps accumulating year after year, thus depleting soil fertility. This coupled with inadequate and imbalanced fertiliser use has further resulted in emergence of secondary and micronutrient deficiencies.Thus, adequate and balanced use of NPK fertilisers as per soil tests is indispensable. This is true for all soils under cropped areas. But,this is not happening. It has resulted in the mining of soil nutrients with resultant loss of fertility. Potassium is the most mined nutrient from soils (Table 5) with the removal of 7 Mt and addition of only 1 Mt (Sharma and Singh, 2012).

**Table 5 :** Nutrient mining from soils of India (Mt)

| **Nutrient Gross balance** | **Addition** | **Removal** | **Balance** |
|---|---|---|---|
| N | 10.9 | 9.6 | 1.3 |
| $P_2O_5$ | 4.2 | 3.7 | 0.5 |
| $K_2O$ | 1.4 | 11.6 | -10.2 |
| Total | 16.5 | 24.9 | -8.4 |
| **Nutrient Net Balance** | | | |
| N | 5.5 | 7.7 | -2.2 |
| $P_2O_5$ | 1.5 | 3.0 | -1.5 |
| $K_2O$ | 1.0 | 7.0 | -6.0 |
| Total | 8.0 | 17.7 | -9.0 |

*Source* : Tandon (2014)

Tiwary (2007) summarised the nutrient balance in soils under crop production in several Indian states. Analysis of the situation in a few states reveal that, in most states the balance for K is strongly negative with excess use of N and sometimes, P (Table 6)

**Table 6 :** Balance sheet of NPK in soils of a few states

| Name of the state | Nutrients Balance ('000t ) in soil over years (1980-2006) | | |
|---|---|---|---|
| | Addition | Removal | Balance |
| **Punjab** | | | |
| N | 1081.1 | 588.6 | 492.5 |
| $P_2O_5$ | 275.4 | 279.3 | -3.9 |
| $K_2O$ | 18.7 | 763.5 | -744.8 |
| **Haryana** | | | |
| N | 847 | 463 | 384 |
| $P_2O_5$ | 253 | 218 | 35 |
| $K_2O$ | 29 | 652 | -623 |
| **Uttar Pradesh** | | | |
| N | 2390 | 1500 | 890 |
| $P_2O_5$ | 739 | 531 | 208 |
| $K_2O$ | 181 | 1972 | -1791 |
| **Madhya Pradesh** | | | |
| N | 519 | 695 | -176 |
| $P_2O_5$ | 344 | 431 | -87 |
| $K_2O$ | 24 | 849 | -825 |
| **Rajasthan** | | | |
| N | 547 | 835 | -288 |
| $P_2O_5$ | 147 | 235 | -88 |
| $K_2O$ | 7 | 1068 | -1061 |

Nutrient balance calculations in most of the cases prior to 2009 were based on the nutrient removal by the crops and addition through fertilisers neglecting contribution from sources other than fertilisers such as organic manures, crop residues and stubbles, irrigation water. Satyanarayana and Tewatia (2009) calculated nutrient balance and mining index in major states of India based on the FAI statistics and information available in literature for the nutrient removal by field, horticultural and plantation crops and additions through fertilisers, organics, irrigation water etc. The trends of nutrient balances for major states presented in Tables 7 to 10, zone wise, reveals that the gap between nutrient use and supply in farming areas will continue to grow wide on account of intensive cropping and therefore, there is need to ensure proper and timely supply of nutrients. They also, pointed out that the nutrient use pattern in majority of the states of India is inadequate and mostly dominated by NP fertilization. The negative balance of K is highly predominant in almost all the states, which imply that the use of K fertilisers is neglected in most cases, K addition through the prevailing level of fertiliser application, practice of manuring and residue recycling are not sufficient to match the K removal by different crops and therefore, sustained efforts are needed to improve K fertiliser consumption through K rich fertilisers and prevent further K mining from soils.

**Table 7 :** Balance sheet of NPK in major states of eastern India

| **State** | **Nutrients ('000t )** | | | **Mining Index (R/A)** |
|---|---|---|---|---|
| | **Addition (A)** | **Removal (R)** | **Balance** | |
| **Assam** | | | | |
| N | 103.4 | 216.1 | -112.8 | 2.1 |
| $P_2O_5$ | 54.6 | 65.1 | -10.5 | 1.2 |
| $K_2O$ | 56.0 | 255.8 | -199.8 | 4.6 |
| Total | 213.9 | 537.1 | -323.1 | 7.9 |
| **Bihar** | | | | |
| N | 929.6 | 452.8 | 476.8 | 0.5 |
| $P_2O_5$ | 191.6 | 203.2 | -11.6 | 1.1 |
| $K_2O$ | 84.4 | 646.4 | -561.9 | 7.7 |
| Total | 1205.6 | 1302.4 | -96.8 | 9.2 |
| **Jharkhand** | | | | |
| N | 89.4 | 122.5 | -33.1 | 1.4 |
| $P_2O_5$ | 45.8 | 59.5 | -13.6 | 1.3 |
| $K_2O$ | 9.8 | 169.6 | -159.9 | 17.4 |
| Total | 145.0 | 351.6 | -206.6 | 20.0 |
| **Odisha** | | | | |
| N | 272.1 | 264.9 | 7.2 | 1.0 |
| $P_2O_5$ | 116.8 | 135.3 | -18.5 | 1.2 |
| $K_2O$ | 63.0 | 383.6 | -320.6 | 6.1 |
| Total | 451.9 | 783.8 | -331.9 | 8.2 |
| **West Bengal** | | | | |
| N | 684.5 | 676.7 | 7.8 | 1.0 |
| $P_2O_5$ | 385.8 | 319.3 | 66.5 | 0.8 |
| $K_2O$ | 304.4 | 972.8 | -668.4 | 3.2 |
| Total | 1374.7 | 1968.8 | -594.1 | 5.0 |

A:- Nutrient additions only through fertilisers

R:- Nutrient removal calculated as per FAI statistics

**Table 8 :** Balance sheet of NPK in major states of northern India

| State | Nutrients Balance ('000t ) | | | Mining Index (R/A) |
|---|---|---|---|---|
| | Addition (A) | Removal (R) | Balance | |
| **Uttar Pradesh (a)** | | | | |
| N | 3180.7 | 1488.3 | 1692.4 | 0.5 |
| $P_2O_5$ | 1023.6 | 591.8 | 431.8 | 0.6 |
| $K_2O$ | 863.1 | 1842.1 | -979.0 | 2.1 |
| Total | 5067.4 | 3922.2 | 1145.2 | 3.2 |
| **Punjab (b)** | | | | |
| N | 1315.5 | 750.2 | 565.2 | 0.6 |
| $P_2O_5$ | 343.9 | 446.3 | -102.4 | 1.3 |
| $K_2O$ | 38.4 | 1022.8 | -984.5 | 26.7 |
| Total | 1697.8 | 2219.3 | -521.6 | 28.5 |
| **Haryana (c)** | | | | |
| N | 520.6 | 489.8 | 30.8 | 0.9 |
| $P_2O_5$ | 65.1 | 220.3 | -155.2 | 3.4 |
| $K_2O$ | 16.6 | 669.1 | -625.5 | 40.4 |
| Total | 602.3 | 1379.2 | -776.9 | 44.7 |

a : Nutrient additions through fertilisers, irrigation water, crop residues and FYM.

b : Nutrient additions through fertilisers only and other sources were negligible.

c : Nutrient additions through fertilisers, 1/3[rd] contribution of nutrients from cow dung and efficiency factors of 0.55, 0.25 and 0.66 were considered for N, P and K, respectively.

**Table 9 :** Balance sheet of NPK in major states of southern India

| State | Nutrients Balance ('000t ) | | | Mining Index (R/A) |
|---|---|---|---|---|
| | Addition (A) | Removal (R) | Balance | |
| **Andhra Pradesh (a)** | | | | |
| N | 786.0 | 505.5 | 280.5 | 0.6 |
| $P_2O_5$ | 187.1 | 232.8 | -45.8 | 1.2 |
| $K_2O$ | 411.3 | 708.6 | -297.3 | 1.7 |
| Total | 1384.4 | 1446.9 | -62.6 | 3.6 |
| **Karnataka (b)** | | | | |
| N | 790.3 | 569.2 | 221.1 | 0.7 |
| $P_2O_5$ | 386.8 | 295.8 | 91.0 | 0.8 |
| $K_2O$ | 330.3 | 734.4 | -404.0 | 2.2 |

*Contd...*

| | | | | |
|---|---|---|---|---|
| Total | 1507.4 | 1599.4 | -92.0 | 3.7 |
| **Tamil Nadu (b)** | | | | |
| N | 543.3 | 494.8 | 48.6 | 0.9 |
| $P_2O_5$ | 228.1 | 230.9 | -2.8 | 1.0 |
| $K_2O$ | 304.2 | 726.1 | -421.9 | 2.4 |
| Total | 1075.7 | 1451.7 | -376.1 | 1.3 |
| **Kerala (b)** | | | | |
| N | 93.3 | 185.6 | -92.3 | 2.0 |
| $P_2O_5$ | 42.7 | 76.9 | -34.2 | 1.8 |
| $K_2O$ | 72.3 | 278.6 | -206.3 | 3.9 |
| Total | 208.3 | 541.1 | -332.8 | 2.6 |

a : Nutrient additions and removal along with efficiency factor and 10% of available potential organic manure considered.

b : Nutrient additions through fertilisers only.

**Table 10 :** Balance sheet of NPK in major states of western India

| State | Nutrients Balance ('000t ) | | | Mining Index (R/A) |
|---|---|---|---|---|
| | Addition (A) | Removal (R) | Balance | |
| **Madhya Pradesh (a)** | | | | |
| N | 934.2 | 1053.2 | -118.9 | 1.1 |
| $P_2O_5$ | 471.0 | 385.8 | 85.2 | 0.8 |
| $K_2O$ | 168.9 | 943.9 | -775.0 | 5.6 |
| Total | 1574.1 | 2328.9 | -808.8 | 7.5 |
| **Gujarat (b)** | | | | |
| N | 1052.6 | 872.1 | 180.5 | 0.8 |
| $P_2O_5$ | 424.5 | 771.7 | -347.2 | 1.8 |
| $K_2O$ | 146.1 | 1137.4 | -991.3 | 7.8 |
| Total | 1623.3 | 2781.2 | -1157.9 | 10.4 |
| **Rajasthan (b)** | | | | |
| N | 705.3 | 825.1 | -119.8 | 1.2 |
| $P_2O_5$ | 260.5 | 371.1 | -110.7 | 1.4 |
| $K_2O$ | 20.9 | 1014.8 | -993.9 | 48.5 |
| Total | 986.7 | 2211.1 | -1224.4 | 51.1 |

*Contd...*

| Maharashtra (b) | | | | |
|---|---|---|---|---|
| N | 1263.5 | 1088.1 | 175.4 | 0.9 |
| $P_2O_5$ | 641.5 | 683.7 | -42.2 | 1.1 |
| $K_2O$ | 420.8 | 1482.9 | -1062.1 | 3.5 |
| Total | 2325.9 | 3254.8 | -928.9 | 5.5 |

a : Nutrient additions through fertilisers and organic manure considered. Contribution of BNF deducted for calculating N removal.

b : Nutrient additions through fertilisers only.

## 4. Soil Fertility and Nutrient Availability trends

Several workers across the country are assessing the fertility status of the soils, which is under crop production. Soil fertility provides information on the status of available plant nutrients needed for plant growth. Tandon (2013, 2014) has provided a valuable account of the current scenario of soil fertility, wherein, he has indicated that 90-95% districts of India are either low or medium in available N and P, while 90% soils of the districts are either medium or high in available K (Table 11)

**Table 11** : Summary of the NPK fertility status of Indian soils over the years (**)

| Year | Nutrient | No. of districts covered | %age of districts | | |
|---|---|---|---|---|---|
| | | | Low | Medium | High |
| 1960s | N | 224 | 52 | 43 | 05 |
| | P | 226 | 47 | 49 | 04 |
| | K | 184 | 19 | 53 | 28 |
| 1997-99 | N | 3.65* | 63 | 26 | 11 |
| | P | 3.65* | 42 | 38 | 20 |
| | K | 3.65* | 13 | 37 | 50 |
| After 2005-06 | N | 500 | 57 | 37 | 06 |
| | P | | 51 | 40 | 09 |
| | K | | 09 | 42 | 49 |

Data for 1960s based on IARI (1980), those for 1997-99 based on Motsara (2002) and for post 2005 on Muralidharu *et al.* (2011).

* No. of samples analysed-3.65 million.

** After Tandon (2013)

Complementary use of chemical fertilisers, organic manures and biofertilisers referred to as integrated nutrient management, is a must to arrest nutrient mining from soils due to cropping and improve/sustain the physical and biological fertility of soil. Long term fertiliser experiments across the country has conclusively shown that compost added in conjunction with recommended fertilisers sustain high crop yields over years without any adverse effect on soil health under various cropping systems in different agro-ecological situations for contrasting soils of India (Table 12).

**Table12 :** Mean grain yield (1972-2009) of crops (t $ha^{-1}$) of predominant cropping systems under long term fertiliser experiments at different centres in India.

| Centre | Crops | Control | N | NP | NPK | NPK+FYM | NPK+Lime |
|---|---|---|---|---|---|---|---|
| New Delhi | Maize | 1.22 | 1.55 | 1.78 | 2.19 | 2.42 | |
| Inceptisols | wheat | 2.38 | 4.08 | 4.52 | 4.52 | 4.87 | |
| | WEY | 3.28 | 5.22 | 5.83 | 6.13 | 6.64 | |
| Ludhiana | Maize | 0.90 | 1.89 | 2.37 | 3.12 | 3.80 | |
| Inceptisols | Wheat | 1.21 | 3.00 | 4.12 | 4.77 | 4.99 | |
| | WEY | 1.87 | 4.39 | 5.86 | 7.05 | 7.77 | |
| Barrackpore | Rice | 1.43 | 2.98 | 3.31 | 3.44 | 3.65 | |
| Inceptisols | Wheat | 0.76 | 1.96 | 2.18 | 2.29 | 2.33 | |
| | WEY | 1.75 | 4.03 | 4.48 | 4.68 | 4.86 | |
| Jabalpur | Soybean | 0.81 | 1.02 | 1.65 | 1.82 | 2.00 | |
| Vertisols | Wheat | 1.24 | 1.67 | 4.07 | 4.42 | 4.85 | |
| | WEY | 2.49 | 3.24 | 6.61 | 7.22 | 7.93 | |
| Pantnagar | Rice | 2.82 | 4.66 | 4.83 | 5.00 | 5.72 | |
| Mollisols | Wheat | 1.43 | 3.59 | 3.76 | 3.77 | 4.49 | |
| | WEY | 3.39 | 6.82 | 7.11 | 7.24 | 8.45 | |
| Palampur | Maize | 0.29 | 0.42 | 2.00 | 3.24 | 4.66 | 4.11 |
| Alfisols | Wheat | 0.38 | 0.37 | 1.64 | 2.29 | 3.10 | 2.85 |
| | WEY | 0.60 | 0.68 | 3.11 | 4.66 | 6.51 | 5.86 |
| Pattambi | Rice | 1.60 | 2.20 | 2.01 | 2.26 | 2.77 | 2.63 |
| Alfisols | Rice | 2.20 | 3.07 | 3.01 | 3.16 | 3.63 | 3.54 |
| | WEY | 2.64 | 3.65 | 3.48 | 3.76 | 4.44 | 4.28 |
| Ranchi | Soybean | 0.61 | 0.29 | 0.87 | 1.50 | 1.87 | 1.80 |
| Alfisols | Wheat | 0.69 | 0.38 | 2.45 | 2.80 | 3.33 | 3.20 |
| | WEY | 1.63 | 0.83 | 3.79 | 5.11 | 6.21 | 5.97 |

Recalculated based on the data of AICRP-LTFE, WEY: Wheat equivalent yield

Legumes should be an integral part of important crop rotations. Emphasis should be on management and recycling of crop residues in soil fertility management. Besides legumes, biofertilisers can be an important component in the nutrient supply system in a cropping system. Some important biofertilisers, their mode of action, host plants and mode of application has been given by Narayanasamy *etal.* (2009) and presented in Table 13.

**Table 13 :** Biofertilisers and their mode of action on host plants

| Organism | Mode of action | Host crops | Method & rate of application | Remarks |
|---|---|---|---|---|
| Rhizobium | symbiotic $N_2$ Fixation | Legumes | seed treatment (20g/kg seed) | residual N left by crops for use by next crop |
| Azotobacter | non-symbiotic $N_2$ fixation | cereals, millet vegetables, cotton | do | benefits crops & helps to control diseases |
| Azospirillum | Associative $N_2$ fixation | maize, barley, sorghum, millet, rice, oats | do | growth promotion |
| Acetobacter | $N_2$ fixation | Sugarcane | sett treatment | |
| Phosphate Solubilisers | solubilisation of P in soil | soil application for all crops | seed treatment (20g/kg seed) can be mixed with rock phosphate | |
| Blue green Algae | non-symbiotic N2 fixation | transplanted rice | soil application (10kg/ha) | improves N nutrition of crop |
| Azolla | Symbiotic $N_2$ Fixation with BGA | Rice | soil application (It dried material/ha) | helps in nitrogen nutrition/ used as Green manure |
| VAM/AM | Association with roots Litchi & other trees | soil application Ornamentals/ sorghum/onion etc. | Soil application | improves mobility esp. P |

Tandon (2011) has provided information on the average content of plant nutrients in organic materials and crop residues (Table 14). Looking into the low soil organic carbon content of our soils, it is important to add organic materials in soils to increase the use efficiency of plant nutrients. It is estimated that about 29Mha of Indian soils are experiencing a decline in soil fertility, which needs to be corrected. Most of the plant biomass is removed from the field to be used as forage, fuel, or building material and stubbles are burnt to facilitate early land preparation for the subsequent crop. Studies have shown that cereal-cereal cropping system, which accelerates nutrient mining, can be made more productive by inclusion of a legume as green manure with positive effects on soil fertility.

**Table 14 :** Average nutrient content of common organic sources

| Sl No. | Organic material | Nutrient content (kg/t of the material) | | | |
|---|---|---|---|---|---|
| | | N | $P_2O_5$ | $K_2O$ | Total |
| 1. | Cattle dung | 3.5 | 1.3 | 1.7 | 6.4 |
| 2. | Sheep & goat droppings | 6.5 | 5.0 | 0.3 | 11.8 |
| 3. | FYM | 7.5 | 2.0 | 5.0 | 14.5 |
| 4. | Rural compost | 7.5 | 2.0 | 5.0 | 14.5 |
| 5. | Urban compost | 17.5 | 10.0 | 15.0 | 42.5 |
| 6. | Water hyacinth compost | 20.0 | 10.0 | 23.0 | 53.0 |
| 7. | Rice straw | 6.1 | 1.8 | 13.8 | 21.7 |
| 8. | Wheat straw | 4.8 | 1.6 | 11.8 | 18.2 |
| 9. | Sugarcane trash | 4.0 | 1.8 | 12.8 | 18.6 |
| 10. | Biogas slurry | 14.1 | 7.2 | 8.4 | 29.4 |
| 11. | Pressmud cake | 21.0 | 44.0 | 8.0 | 73.0 |

The problems of soil fertility losses over the years have shown a gradual increase, primarily due to our poor understanding of the important soil processes. To illustrate this, the case of soil potassium depletion may be cited. Sanyal (2014) has analysed the situation of K in the Indo-Gangetic Plains under rice-wheat, rice-rice and rice-potato crops. Alluvial soils of the region rich in micaceous minerals, have relatively high K-supplying power. Decontrol of Potassic fertiliser prices (and enhanced costs) and often poor response of crops to K have prompted the farmers to use low or no K-fertiliser in crops. This increasing omission of K has led to its mining from native sources. Further, removal of crop residues from the fields has enhanced K-mining from the non-exchangeable pool. This goes unnoticed in the routine soil tests for plant available K (water soluble + exchangeable), while it is known that non-exchangeable forms of K in soil is in dynamic equilibrium with exchangeable K. Such unnoticed K depletion in the long run is likely to disturb the equilibrium among different K fractions, and building up of the soil K fertility in such cases will be very difficult, time consuming and expensive, which the farmers will find difficult to afford.

Some of the measures, which can check nutrient mining from soils are: adequate and balanced nutrient use, improved ratios of N, P, and K, integrated plant nutrient management, crop diversification, residue recycling and conservation agriculture. In Punjab, most of the rice straw produced is burnt in open fields and wheat straw is converted to *Bhusa* for animal feeding using mechanical equipments. This creates pollution problems due to emission of toxic and green house gases, besides loss of organic matter and nutrients. In some areas, green manuring of Dhaincha (*Sesbania aculata*), moong, cowpea, sunhemp after harvest of wheat maintains soil fertility especially N and organic matter. Use of biofertilisers, such as BGA, Rhizobial and Azotobacter cultures need to be promoted on a large scale along with fertilisers and organic manures for better results. Inclusion of legumes in cropping systems improves the biological fertility of soils. Preparation of inventory of organic sources and recyclable wastes under different farming situations is a pre-requisite for generation of location specific IPNS technologies and their wider adoption. Crop diversification is becoming an increasingly important area of agronomic research for improving nutrient use efficiency and judicious utilization of natural resources. There is an urgent need to generate systematic information on nutrient balances with respect to six important nutrients i.e.N, P, K, S, Zn and B, deficiency of which have emerged in many cultivated soils.

Increased mining of soil Potassium without caring for its replacement has resulted in poor soil K status in many soils. Widening of N,P,K, ratios in nutrient use for crop production in many states has complicated the issue.Recent work of Pattanayak *etal.* (2008) has shown that grain yield of crops increased as the NPK ratio was narrowed from 6.4:3.8:1 to 1.6:0.9:1. Some of the important observations on K-use in Indian soils (Prasad, 2009) are as follows:

- Red and lateritic soils spread over eastern, north-eastern and southern India, which have rice as a dominant crop and are poor in all forms of K and total K, need adequate fertilisation beyond the ratio of 4:2:1.
- Black soils, where sorghum, soybean, pigeonpea and groundnut are the dominant crops rich in exchangeable K, but poor in non-exchangeable and total K, potassium use need to be increased.
- Alluvial soils of north India with rice-wheat cropping system are rich in non-exchangeable K and total K in surface soils as well as subsoil, but there are pockets where K mining is high. In such areas, K fertilisation needs to be improved.
- An all out effort needs to be made to meet part of the K needs of crops from organic manures and crop residues (Table 15).

**Table 15 :** NPK content (kg/tonne) of some crop residues

| Crop residues | N | $P_2O_5$ | $K_2O$ | Total |
|---|---|---|---|---|
| Rice | 6.1 | 1.8 | 13.8 | 21.7 |
| Wheat | 4.8 | 1.6 | 11.8 | 18.2 |
| Sorghum | 5.2 | 2.3 | 13.4 | 20.9 |
| Maize | 5.2 | 1.8 | 13.5 | 20.5 |
| Pearl millet | 4.5 | 1.6 | 11.4 | 17.5 |
| Pulses | 12.9 | 3.6 | 16.4 | 32.9 |
| Oilseeds | 8.0 | 2.1 | 9.3 | 19.4 |
| Groundnut | 16.0 | 2.3 | 13.7 | 32.0 |
| Sugarcane | 4.0 | 1.8 | 12.8 | 18.6 |

Average values compiled from different sources (Prasad, 2009)

The deficiency of Sulphur and micronutrients are on the rise in India.This is primarily due to intensive cropping, use of S-free fertilisers and poor organic matter content of soils. Leaching of soluble sulphur in high rainfall areas is a problem leading to its deficiency. About 240 districts have reported S-deficiency problems in soils (Singh, 2006). Based on the analysis of 60,000 soil samples, 41% soils were reported to be deficient in available Sulphur (Singh,2001). During 1997 to 2006, 49000 soil samples from different regions were analysed for the status of $CaCl_2$ extractable S under TSI-FAI-IFA Sulphur Project. It was found that 46% soils were deficient (Tewatia *et al.*, 2006) and another 30% soils were potentially deficient (Table 16).

**Table 16 :** Sulphur status of Indian soils

| Region | No. of samples | % of samples in category | | |
|---|---|---|---|---|
| | | Low | Medium | High |
| Northern | 15323 | 44 | 30 | 26 |
| Western | 12474 | 45 | 30 | 25 |
| Eastern | 10108 | 35 | 33 | 32 |
| Southern | 11289 | 63 | 26 | 11 |
| All India | 49194 | 46 | 30 | 24 |

The states of India most deficient in S are Himachal Pradesh (84%), Kerala (81%), Rajasthan (65%), Andhra Pradesh (56%), and Jharkhand (51%). The existing gap between addition (0.8 Mt) and removal (1.8Mt) is over 1 Mt per annum, which is increasing with time. Farmers still are not aware of the role of micronutrients in crop production and how much harm these can do to a standing crop. As per the current status (Table 17), cationic micronutrients in Indian soils (Shukla, 2012; Tandon, 2013), is estimated to be deficient in 42% (Zn), 4% (Cu),

15% (Fe), and 6% (Mn). For, anionic micronutrient like Boron,it is over 32%. Zn deficiency is common in many intensively cropped soils,especially the rice based systems. In winter maize,this is common in alluvial calcareous soils. Boron deficiency is common in red and lateritic soils,especially in vegetable based systems. In coarse textured soils,low in organic matter, B deficiency is common.Iron deficiency has been observed in upland soils,especially in rice. Deficiency of Mn is appearing in wheat crop, when it is grown after rice in coarse textured alkaline soils with low organic matter contents.

It is difficult to sustain the desired fertility status of soil micronutrients under intensive cropping. Soil pH controls the availability of micronutrients, such as Zn and Cu. Boron availability is controlled by soil organic matter status. Iron and manganese availability is governed by soil physiography, upland or lowland and aerobic or waterlogged situations. Soil texture (light or heavy) influences the forms of micronutrients in soil (water soluble,exchangeable,fixed etc.). Further, when micronutrient fertilisers are applied to soils,bulk of it remains in soil, while only 2 to 5% of the nutrient is taken up by plants. In view of this low nutrient use efficiency of micronutrient fertilisers by crops, soil + foliar application is preferred. This implies that effective management of soil micronutrients for crop production under different soils and cropping systems require an in-depth knowledge of the dynamics of these nutrients in the soil-plant system.

**Table 17 :** Status of deficiency of micronutrients in Indian Soils

| State | % samples deficient | | | | |
|---|---|---|---|---|---|
| | **Zn** | **Cu** | **Fe** | **Mn** | **B** |
| Andhra Pradesh | 44.3 | 0.24 | 1.82 | 1.14 | 5.6 |
| Assam | 23.8 | 0.31 | 0.00 | 0.00 | 16.5 |
| Bihar | 55.7 | 4.01 | 8.37 | 6.53 | 30.5 |
| Chhattisgarh | 29.5 | 0.20 | 1.87 | 0.00 | 43.6 |
| Gujarat | 31.2 | 0.13 | 15.90 | 9.15 | 2.8 |
| Haryana | 16.9 | 2.02 | 22.41 | 8.60 | 2.1 |
| Jharkhand | 9.8 | 5.60 | 0.89 | 1.05 | 65.9 |
| Maharashtra | 47.7 | 4.32 | 17.32 | 1.12 | 9.8 |
| Madhya Pradesh | 46.7 | 0.67 | 8.09 | 0.41 | 22.1 |
| Odisha | 11.3 | 1.25 | 0.08 | 0.00 | 66.5 |
| Punjab | 24.6 | 2.69 | 22.11 | 19.65 | 10.8 |
| Tamil Nadu | 68.6 | 32.91 | 16.10 | 10.17 | 12.1 |
| Uttar Pradesh | 55.7 | 6.04 | 8.92 | 8.29 | 24.3 |
| Uttarakhand | 8.6 | 1.14 | 1.84 | 0.79 | 3.9 |
| West Bengal | 41.6 | 12.16 | 0.93 | 2.06 | 68.4 |
| All India | 42.7 | 4.23 | 14.91 | 5.96 | 32.6 |

*Source* : Tandon(2013)

## 5. Carbon Sequestration in Soils

There is a growing concern that increasing levels of $CO_2$ in the atmosphere will change climate, making our earth warmer with subsequent increase in extreme weather events. Soil carbon sequestration is the process of transferring $CO_2$ from the atmosphere into the soil through crop residues and other organic solids, and in a form that is not immediately remitted. This transfer of carbon helps in off-setting emissions from fossil fuel combustion and other carbon-emitting activities, while enhancing soil quality and sustainability of crop production. Soil carbon sequestration can be accomplished by management systems that add high amounts of biomass to soil, cause minimal soil disturbance, conserve soil and water, improve soil aggregation and enhance activities of soil organisms.

Rough estimates say that principal C reservoir in the earth's surface is not terrestrial or marine biomass, but soil organic matter, the latter amounting to about 1500 to 2000 Pg (1Pg=petagram=1 billion tonne) soil organic C (Batjes, 1996; Post *et al.*,1982). Soil organic pool thus represents more than twice the C in living vegetation (around 560 Pg) or that in the atmosphere (750 Pg).

According to Rattan Lal, Director of Ohio State University Carbon Management and Sequestration Centre, the world's cultivated soils have lost 50 to 70% of their original carbon stock, much of which has oxidized upon exposure to air to become $CO_2$. He says that such a destructive process can be made regenerative by proper agricultural management practices, which can turn back the carbon stock, reducing atmospheric $CO_2$, while boosting soil productivity and increasing resilience to floods and drought. Priority should be to restore degraded and eroded lands, avoid deforestation and farming of peatlands, which are major reservoir of Carbon and are easily decomposed upon drainage and cultivation. As per estimate, restoring soils of degraded and desert ecosystems has the potential to store in world soils an additional 1 to 3 billion tons of Carbon annually, equivalent to roughly 3 to 11 billion tons of $CO_2$ emissions (annual $CO_2$ emission from fossil fuel burning is roughly 32 billion tons).

Soil organic carbon is the single most important parameter affecting soil quality. Soil management practices like residue incorporation, manuring, reduced tillage, mulching can play important roles in sequestering carbon in sols. Increase in SOC by 1Mg/ha can result in an increase in grain yield by 30-50 kg/ha of rice, wheat, millets, beans and 100-140 kg/ha of maize and sorghum with appropriate soil and crop management technologies (Lal,2005). In India, restoration of degraded soils and adoption of BMPs have good scope for Carbon sequestration. This has been amply demonstrated by research carried out under ICAR AP-Cess fund Project on "Carbon Sequestration Strategies in rainfed production systems in India" (Srinivasa Rao, 2009). As per Lal (2004) potential of carbon sequestration in degraded soils of India, if restoration is undertaken is 9.8 to 13.9 Tg/year (Table 18).

**Table 18 :** Extent of soil degradation in India and potential of carbon sequestration in these soils, if restoration is undertaken

| Degradation | Area (Mha) | SOC sequestration potential (Tg/Y) |
|---|---|---|
| Water erosion | 32.8 | 2.6-3.9 |
| Wind erosion | 10.8 | 0.4-0.7 |
| Soil fertility decline | 29.4 | 3.5-4.4 |
| Water logging | 3.1 | 0.1-0.2 |
| Salinization | 4.1 | 0.5-0.6 |
| Desertification | 68.1 | 2.7-4.1 |
| Total | 148.3 | 9.8-13.9 |

*Source* : Lal (2014)

## 6. Soil Organic Carbon and Fertility

Soil organic carbon(SOC) is a major factor contributing to the buffering capacity of soils and soil's ability to hold cations and anions and prevent losses of plant nutrients. It is a source of plant nutrients, especially N,P,S, and micronutrients. The nutrients present in soil organic matter are released to plants upon mineralisation, which is a slow process. It supports a wide variety of microorganisms. It resists changes in soil pH due to its buffering properties. SOC is the most important indicator of soil quality. There is not much realization among farmers on the use of well decomposed organic manures along with chemical fertilisers to build up soil health. This has led to a decline in soil health in many parts of the country, especially under continuously cropped situations. Soil degradation and loss of top soil, removal of crop residues (roots, straw, stubbles) after crop harvest, have accentuated the problem. The status of SOC in Indian soils is generally very low (Table 19).

**Table 19 :** Organic carbon content of major soil groups of India

| Soil | Organic C (%) |
|---|---|
| Black soil | 0.3-1.2 |
| Red and Lateritic soils | 0.3-0.7 |
| Alluvial soils | 0.3-1.1 |
| Hill and mountain soils | 4.0-8.0 |
| Desert soils | 0.2-0.6 |
| Coastal alluvial soils | 0.5-0.9 |

Long term field trials across the country have demonstrated that SOC increased in NPK+FYM treatments. Fertiliser N, NP application over the years brought about a decline in SOC. NPK application without Compost did not improve the SOC. These results clearly point out (Table 20) the need of Integrated plant nutrient management in restoring/improving the SOC levels for sustained production of crops (Manna *et al.* 2003).

**Table 20 :** Effect of long term fertiliser use on Soil organic carbon in India

| Site/soil | Cropping System | SOC (%) | | | |
|---|---|---|---|---|---|
| | | Initial | No fertiliser | NPK | NPK+FYM |
| Bhubaneshwar (Inceptisol) | Rice-Rice | 0.27 | 0.41 | 0.59 | 0.76 |
| Pantnagar (Mollisol) | Rice-wheat | 1.48 | 0.50 | 0.95 | 1.51 |
| Faizabad (Inceptisol) | Rice-Wheat | 0.37 | 0.19 | 0.40 | 0.50 |
| Barrackpore (Inceptisol) | Rice-Wheat-jute | 0.71 | 0.51 | 0.72 | 0.76 |
| Jabalpur (Chromustert) | Soyabean-wheat-Maize fodder | 0.57 | 0.53 | 0.62 | 0.97 |
| Coimbatore (vertic Ustocrept) | Fingermillet-maize-cowpea | 0.30 | 0.44 | 0.50 | 0.64 |
| Delhi (Inceptisol) | Maize-Wheat | 0.44 | 0.32 | 0.47 | 0.56 |
| Palampur (Alfisol) | Maize-Wheat | 0.79 | 0.62 | 0.83 | 1.20 |
| Karnal (Inceptisol) | Rice-wheat-Fallow | 0.23 | 0.30 | 0.32 | 0.35 |
| Nagpur (Vertisol) | Rice-Rice | 0.41 | – | – | 0.55 |
| Trivendrum (Ultisol) | Cassava | 0.70 | 0.26 | 0.60 | 0.98 |

*Source* : Singh (2012) from AICRP-LTFE Reports

Mandal *et al.* (2007) observed that crops and cropping systems play an important role in maintaining SOC stocks as the quantity and quality of residues of these crops that are returned to soils, vary greatly affecting their turnover or residence time in soil (Table 21)

**Table 21 :** Annual crop residue C inputs (Mg C /ha/year) in some rice based cropping systems

| System | Treatments | | |
|---|---|---|---|
| | **Unfertilised** | **NPK** | **NPK+ organic manure** |
| Rice-Mustard-Sesame | 1.88 | 2.76 | 3.75 |
| Rice-Wheat-Fallow | 1.82 | 3.33 | 3.97 |
| Rice-Fallow-Berseem | 2.45 | 3.17 | 4.16 |
| Rice-Wheat-Jute | 2.58 | 5.08 | 6.17 |
| Rice-Fallow-Jute | 2.58 | 3.56 | 4.30 |
| Average | 2.26 | 3.58 | 4.47 |

*Source* : Mandal *et al.*, 2007

## 7.0 Nutrient Management Strategies

Efficient and balanced use of fertilisers in crops is essential. At present, there is distortion in N, P and K, consumption ratio across zones and states. The N:P:K ratio on all India basis confirms this observation (Table 22).

**Table 22 :** N:P:K ratio in nutrient use in India

| Year | N | P | K |
|---|---|---|---|
| 2007-08 | 5.5 | 2.1 | 1.0 |
| 2008-09 | 4.6 | 2.0 | 1.0 |
| 2009-10 | 4.3 | 2.0 | 1.0 |
| 2010-11 | 4.7 | 2.3 | 1.0 |
| 2011-12 | 6.7 | 3.1 | 1.0 |

The ratio of K:P in Haryana is around 1:11, while it is 1:7 in Punjab. Similarly, K:N ratio is 1:32 in Haryana and 1:23 in Punjab, this is not acceptable and is adversely affecting crop yields. Nutrient based subsidy (NBS) has made P and K fertilisers commercially viable with increased availability. But, Urea accounting for about 50% of the total fertiliser application has been left out of NBS and the Industry has no control over its production and supply. Due to high global prices, the production cost of P and K fertilisers have soared during the past 3-4 years. The cumulative effect of this has been a wider NPK use ratio, resulting in imbalance nutrient use. Recent decisions of the Central government for direct transfer of NBS to farmers and monitoring movement of fertilisers upto retail points will probably help to tide over the situation. At present, the micro-level situation is not good and need immediate attention.

There is need to include secondary and micronutrient fertilisers in the schedule of nutrient application. For both yield and quality of crops, it is essential to apply fertilisers, organic manure and biofertilisers in right proportions.

Data presented in Table 23 show a steep rise in consumption of fertilisers/ha in different states.This resulted in substantial increase in foodgrain productivity. But, the response of fertilisers varied considerably among soils/crops/management conditions.

**Table 23 :** Growth in per unit fertiliser consumption and foodgrain productivity (kg/ha) in different states of India

| State | 1980-81 | 2006-07 | % growth over 1980-81 | | | |
|---|---|---|---|---|---|---|
| | Fertiliser consumption | Foodgrain productivity | Fertiliser consumption | Foodgrain productivity | Fertiliser consumption | Foodgrain productivity |
| A.P. | 45.9 | 1147 | 185.9 | 2231 | 301 | 94.5 |
| Assam | 2.8 | 1073 | 54.6 | 1286 | 1850 | 19.8 |
| Gujarat | 34.4 | 1001 | 124.6 | 1423 | 262.2 | 42.2 |
| H.P. | 17.4 | 1026 | 52.1 | 1714 | 194.2 | 67.0 |
| Haryana | 42.2 | 1597 | 173.0 | 3393 | 304 | 112.5 |
| J&K | 21.4 | 1559 | 78.9 | 1734 | 268.7 | 11.2 |
| Karnataka | 34.4 | 1023 | 114.1 | 1289 | 321.7 | 26.0 |
| Kerala | 31.1 | 1587 | 70.0 | 2331 | 125.1 | 46.9 |
| M.P | 9.2 | 688 | 61.5 | 1167 | 568.5 | 69.2 |
| Maharashtra | 21.2 | 693 | 110.1 | 940 | 419.3 | 35.6 |
| Odisha | 9.6 | 870 | 46.2 | 1359 | 381.2 | 56.2 |
| Punjab | 117.9 | 2458 | 209.2 | 4017 | 77.4 | 63.4 |
| Rajasthan | 8.0 | 531 | 43.3 | 1119 | 441.2 | 110.7 |
| TamilNadu | 63.2 | 1340 | 186.5 | 2610 | 195.1 | 94.8 |
| U.P. | 49.4 | 1206 | 148.4 | 2057 | 200.4 | 70.6 |
| West Bengal | 35.9 | 1358 | 143.2 | 2511 | 298.9 | 84.9 |

*Source* : Fertiliser Statistics. The Fertiliser Association of India, New Delhi, India (1994-2009)

In eastern states, for example, the fertiliser use in crops is much less compared to that in northern, western and southern India (Table 24). This has also resulted in low productivity of crops in many states,such as Jharkhand, Odisha, Assam, Tripura, Manipur etc.

**Table 24 :** Consumption of N, $P_2O5$ and $K_2O$(Kg/ha) in different Eastern States of India

| State | N | $P_2O_5$ | $K_2O$ | Total |
|---|---|---|---|---|
| Assam | 32.8 | 13.0 | 17.4 | 63.2 |
| Bihar | 111.3 | 30.8 | 20.9 | 163.0 |
| Jharkhand | 34 | 19 | 7 | 60 |
| Odisha | 33.7 | 17.1 | 9.1 | 59.9 |
| West Bengal | 74.5 | 47.7 | 45.6 | 167.8 |
| Manipur | 37 | 3 | 1 | 41 |
| Tripura | 18 | 7 | 7 | 32.0 |

For an efficient nutrient management strategy in crops and cropping systems in soils of India, some radical changes are required for a turn around. Some of these are discussed below:

### *Rainfed Crops*

About 58% of the 141Mha cultivated area in India is under rainfed and dryland farming. These areas provide 44% of the total foodgrain production including 75% of pulses and over 90% of sorghum, millet and groundnut (Srinivasa Rao *et al.* 2013). Low soil fertility with low crop productivity are characteristic features of such areas mainly due to poor soil organic carbon, poor soil moisture and risk-prone agriculture. Lack of response of crops to added fertilisers in rainfed areas is a major issue. Farmers are generally not convinced on the probable benefits from nutrient use. Research has shown that combined application of organic manures and fertilisers has a synergistic effect on crop performance, besides improving soil physical and biological environment. Potash application is also highly beneficial in imparting resistance to crops for withstanding drought stress. Practices, such as application of vermicompost, inclusion of legumes in cropping systems,minimum soil disturbance, green leaf manuring, residue management, water resource management etc. need to be promoted with large scale farmers'participation to increase farm productivity and profitability.

### *Intensively Cropped Areas*

The rice-wheat cropping system in the Indo-Gangetic Plains cover an area of around 10Mha.The states of Punjab and Haryana contribute 20% to the total foodgrain production of the country and 50 & 85% of the Government procurement of rice and wheat, respectively. The indiscriminate use or rather misuse, of natural resources,especially water has led to groundwater pollution as well as groundwater depletion.This region requires focussed attention in order to sustain high levels of agriculture production to sustain food security of the population. Some of the important issues for this region are:

- Declining response of crops to per unit of nutrient use
- Increasing deficiency of K, S, Zn, Fe, Mn etc
- Rapidly declining water table in some areas and increasing water table in others
- Increase in problem of salinization
- Deteriorating water quality

Proper attention on the above issues and widespread adoption of some promising technologies in the area will be desirable to solve the problems. Some such technologies are:laser land levelling, zero or minimum tillage, bed planting, surface seeding for wheat and SRI method of rice cultivation, dry seeding of rice, weed management, use of leaf colour chart for N management for rice. Motivating farmers in the above direction will take care of the deficiencies and will ensure high productivity of both rice and wheat.

***Export Oriented Crops***

This group of crops and their processed products can bring about a major shift in Indian agriculture. Crops, such as vegetables, fruits, spices, plantation crops, need special mention in this regard. Out of about 21.2 Mha under horticulture, 14.7 Mha is under fruits and vegetables, plantation and spices crops cover about 5.7 Mha. Nutrient removal by fruit crops is very high, but it is hardly realised by farmers and extension workers in rural areas. A banana crop yielding about 40t/ha removes about 294kgN, 67kg $P_2O_5$ and 1140 kg$K_2O$. A grape crop yielding 30t/ha removes about 230kgN, 83 kg $P_2O_5$ and 356 kg $K_2O$. Nutrient removal by vegetables is also fairly high. A cabbage crop of about 90 days duration yielding 75t/ha removes 410 kg N, 75 kg $P_2O_5$ and 460 kg $K_2O$, while a cauliflower crop yielding 40t/ha removes 230 kg N, 96 kg $P_2O_5$ and 325 kg $K_2O$.

Nutrients, thus extracted from soil must be replenished to sustain soil health and prevent plant nutrient imbalances. It is estimated that N, P, K, requirements for the horticulture sector is around 3.7 Mt, out of about 28Mt of total nutrient use in Indian agriculture. Use of organic manures in these crops is high, but the quality of such manures continue to be poor. This is primarily due to lack of awareness among the farming community and improper field level trainings imparted to them. The technology of enriched composting (use of phosphates, bio-inoculants in compost pits) has not reached the farmers. Use of Azotobacter,VAM or P-solubilisers in compost pits for improving compost quality can help in crop quality enhancement. Fertigation, to improve the efficiency of nutrients as well as water and minimizing nutrient losses is of great relevance in horticultural crops. Nutrient supply in crops needs to be closely linked to such systems to improve the economy of nutrient use in crops.

The productivity of mangoes could be enhanced through wider adoption of improved agro-techniques,such as high density planting, INM, drip and fertigation,clean cultivation etc. Micronutrients, such as Zn, B, Mo, Fe in many cases need to be applied as soil, foliar or a combination of the two to boost yield as well as quality. INM to maximize nutrient use efficiency in spices and tuber crops as well as for controlling the emerging deficiencies of S and Zn will be necessary.

Pulses productivity (average,600-700 kg/ha, with potential of 1.5 to 3t/ha) is extremely low across soils due to their cultivation in marginal and sub-marginal lands. Poor soil fertility with limited use of essential plant nutrients, weed infestation and moisture stress at critical growth stages have resulted in poor yield levels over the last 3 to 4 decades. It is estimated that one tonne of biomass production in pulses, require about 30-50kgN, 2-7kgP, 12-30kg K, 3-10kgCa, 1-5kgMg, 1-3kg S, 200-500g Mn, 5gB, 1gCu and 0.5gMo (Ahlawat & Ali,1993). Thus, to augment productivity, adequate, balanced and integrated nutrient use is essential. 30-40kg $P_2O_5$/ha in rainfed areas and 40-60kg $P_2O_5$in irrigated areas with 20-40$K_2O$/ha is necessary for higher pulses production. Sulphur application@20kg/ha, liming acid soils for growing pulses, 15-20kg/ha $ZnSO_4$,10kg/ha borax and 1kg/ha chelated iron is necessary as per soil tests. Dual inoculation of PSB and Rhizobium has been found effective. Such need based input use in pulse growing soils will bring about improvement in quantity and quality of production.

In oilseed production, the productivity enhancement through balanced input use must receive major thrust. Oilseeds are energy-rich crops and thus,the requirement of N,P,K,S and micronutrients is high.There is a serious concern about deficiencies of secondary and micronutrients, especially S, B, and Zn in large areas under oilseed production. S application alongwith NPK can increase the oil content in seeds by 2-3% in sesame and safflower, 3-4% in sunflower, 4-5% in linseed, 4-7% in soybean and 4-9% in groundnut (Hegde and Sudhakarbabu, 2004).

Oilseed crops are important sources of energy in human diet, that helps to overcome calorie malnutrition. This calls for balanced and integrated use of plant nutrients based on the status of soil fertility to sustain and increase the protein and oil contents. Organic manures and residue management will make oilseed production sustainable in the long run.

## References

Ahlawat, I.P.S. and Ali, M (1993) Fertiliser management in Food crops (HLS Tandon ed.) F.D.C.O., New Delhi, pp 114-138.

Anonymous (2013). Progress reports of AICRP-LTFE, Ranchi centre.

Bhattacharyya, T., Sarkar, D., Sehgal, J. L., Velayutham, M., Gajbhiye, K. S., Nagar, A. P. And Nimkhedkar, S. S. (2009) Soil taxonomic database of India and the states (1:250000 scale) NBSSLUP Publ. 143, National Bureau of Soil Survey and Land Use Planning, Nagpur. pp 266.

Govinda Rajan, S.V. (1971) Soil map of India. In 'Review of Soil research in India' (J.S. Kanwar and S.P. Raychaudhuri, eds.) Indian Society of Soil science, New Delhi.

Govinda Rajan, S.V. and Rao Gopal, H.G.(1978) Studies on Soils of India. Vikas Publ. House Pvt. Ltd., New Delhi.

Hegde, D.M. and Sudhakarababu, S.N. (2004) Balanced fertilisation for nutritional quality in oilseeds. Fert. News. 49(4) pp 57-62.

Lal, R. (2004) Soil Carbon sequestration impacts on global climate change for food security. Science 304, pp 1623-1627.

Mandal, B., Majumdar, B., Bandopadhyay, P.K. (2007) The potential of cropping systems and soil amendments for carbon sequestration in soils under long term experiments in sub-tropical India. Global Change Biology 13, pp 357-369.

Manna etal.(2012) Compost Handbook, F.D.C.O., New Delhi pp 1-132.

Motsara, M.R.(2002) Available nitrogen, Phosphorus and potassium status of Indian Soils as depicted by soil fertility maps. Fert. News 47,15-21.

Narayanasamy, G., Arora, B.R. and Khanna, S.S. (2009) Fertilisers, manures and biofertilisers. In 'Fundamentals of Soil Science', Indian Society of Soil Science Publ., pp 579-622.

NBSS & LUP (2008) Harmonization of degraded lands/wastelands Data sets of India-a GIS based approach, NBSS & LUP (ICAR), Nagpur.

Pattanayak, S.K., Mukhi, A., and Majumdar, K. (2008). Better Crops.92(2) pp8-9.

Prasad, Rajendra (2009) Potassium fertilisation recommendation for crops –Need re-thinking. Indian Journal of Fertilisers 5 (8), pp 31-33.

Sahoo, A.K. and Sarkar, Dipak (2013) Genesis and Classification of Acid Soils of India. In, A K Sarkar ed.Publ.' Acid soils-their chemistry and management' New India Publ. Agency, New Delhi, pp 49-104.

Sanyal, S.K. (2014) Nutrient mining in Agriculture-an issue of concern. Indian Society of Soil Science Newsletter 36, pp 1-3.

Sarkar, A.K.(2013) Acid soils-their chemistry and management. New India Publishing Agency, New Delhi.

Satish Chander (2010) Frank notes: Soil health issues and strategies. Indian Journal of Fert. 6(7) 8-11.

Satyanarayana, T. And Tewatia, R. K. (2009) State wise approach to crop nutrient balance in India. In: Proceedings IPI-OUAT-IPNI International Symposium Brar, M. S. and Mukhopadhyay, S. S. (Eds) pp 467-485.

Sehgal, J.L., Saxena, R.K., and Vadivelu, S (1987) Field Manual-Soil Resource Mapping of different states of India. Tech. Bull. No.13, NBSS & LUP Publ.13.

Sharda V.N., Dogra, Pradeep and Chandra Prakash (2009) Assessment of production losses due to water erosion in rainfed areas of India. Journal of Soil and water Conservation. IOWA, USA.

Sharma, P.D., and Singh, M.V.(2012) State of health of Indian Soils. Platinum Jubilee Symp. Proc. Indian Society of Soil Science, New Delhi, pp 191-213.

Shukla, N.D. (2010) Impact of fertiliser on Indian agriculture and rural prosperity. Indian J of Fertilisers 6(7): 12-27.

Singh, M.V. (2001) Importance of Sulphur in balanced fertiliser use in India. Fert. News 46,13-35.

Singh, M.V. (2006) Micro-, secondary and pollutant elements Research in India. Indian Institite of Soil Science, Bhopal Publ. pp 1-80.

Soil Survey Staff (1999) Soil taxonomy: A basic system of soil classification for making and interpreting soil survey. 2nd edition. NRCS, agriculturehandbook, Washington DC,USA.

Srinivasa Rao, Ch, Ravindra Charry, G. Venkateswarlu, B.,Vittal, KPR, Prasad, JVNS, Kundu, Sumanta, Singh, SR., Gajanan, GN, Sharma, RA, Patel, JJ, Deshpande, AN & Balaguravaiah, G (2009) Carbon sequestration strategies in rainfed Production Systems of india, CRIDA, ICAR, Hyderabad-500059.

Srinivasa Rao, Ch. *et al.* (2009) Carbon stocks in different soil types under diverse rainfed production systems in tropical India. Commn. Soil Sci.Pl. Anal. 40:2338-2356.

Subba R Rao, A., Patra, D.D., and Biswas, A.K.(2009) Soil health and Nutritional security-role of primary nutrients. Platinum Jubilee Symposium Proc. Indian Society of Soil Science, New Delhi, pp 214-228.

Tandon, HLS (2007) Soil Nutrient Balance Sheets in India. Better Crops:15-18.

Tandon, HLS (2011) Biofertilsers and Organic Fertilisers Sourcebook cum Dictionary. F.D.C.O., New Delhi, India pp1-150.

Tandon, HLS (2013) Soil Health management:productivity-sustainability-resource management. F.D.C.O., New Delhi, India, pp1-204.

Tandon, HLS ed. (2014) Soil testing for balanced fertilisation: technology-application-problems-solutions. F.D.C.O., New Delhi, India, pp1-170.

Tewatia, R.K., Choudhary, R.S., & Kalwe, S.P. (2006) TSI-FAI-IFA Sulphur project-salient findings. In Proc. TSI-FAI-IFA Symp. cum Workshop on S in Balanced Fertilisation (RK Tewatiaetal.ed.) Oct.4-5, New Delhi pp 15-25.

Tiwari, K.N. (2008) Breaking yield stagnation through balanced fertilisation in Rajasthan, MP, Punjab, HP, Haryana & UP.I.P.N.I-India Programme, Gurgaon (Haryana).

Velayutham, M and Bhattacharya, T (2000) In chapter I' Soil Resource Management'of International Conference on Managing natural resources for sustaining Agril. Production in the 21stCentury, New Delhi Publ. (Ed.Yadav, JSP and Singh, G.B.) pp 3-129.

❑❑❑

CHAPTER - 4

# Soil Analysis

**P. Mahapatra**

## A. Soil Sampling

Soil is a store house of several nutrients. These are present in available and reserve forms and a dynamic equilibrium always exists between these forms of nutrients in soil. If the available form of a given nutrient element is depleted by crop uptake or by any other means, there will be immediate replenishment from the reserve source. In this context, it is also appropriate to note that soil is not an eternal source of plant nutrients. In a bid to produce more, soils are exploited very badly and they are gradually becoming deficient in several nutrients. There are several soil related constraints coming in the way of achieving sustainability in crop production.

### *1. Physical Constraints*

Coarse texture, deterioration in soil structure, surface crusting, soil compaction, poor permeability, sub- soil pans, poor infiltration, waterlogging and drainage, soil erosion under shifting cultivation.

### *2. Chemical Constraints*

Soil acidification, salinisation in irrigated agriculture and drylands, soil fertility interactions, elemental toxicities arising from build up of heavy metals etc.

### *3. Biological Constraints*

Decline in microbial biomass, soil borne diseases and decline in quality and quantity of organic matter.

To overcome the above mentioned problems and to provide a conducive environment for growth and development of plants, adoption of suitable soil management and conservation practices, amelioration of problem soils, improving soil conditions for achieving higher nutrient use efficiency, choosing crop species to exploit the reserve forms of nutrients, choosing appropriate cultural practices and cropping systems for maximizng utilization of available nutrients, growing legumes and tree crops in the cropping systems and microbial inoculantion to mobilize unavailable nutrients for crop production, need greater attention.

Soil analysis and interpretation of soil testing data for management recommendations is valid only if the test data are accurate. Otherwise, soil test results and its interpretation are inappropriate, misleading and costly for farmers, who follow recommendations based on such invalid data. A soil testing programme consists of four phases:

i. soil sampling;

ii. sample analysis (testing)

iii. soil-test data interpretation

iv. soil management recommendation based on the above

The care and accuracy of conducting each of the first three phases, determine the validity of the fourth phase. Soil sampling is regarded as the most important step in a soil testing programme, because the error due to soil sampling confers error and invalidity on the soil analysis result, even when the analysis is carefully and accurately conducted. Therefore, quality of the soil sample determines, to a great extent, whether or not the soil-test result is useful. A soil management recommendation based on soil analysis is good only if the soil sample analyzed is representative of the field. The quality of the soil sample depends on the care taken in collecting, handling and preparing the sample for analysis.

**Why To Sample Soil?**

It is essential that a soil sample be truly representative of the field, which it is supposed to represent, otherwise, it will lead to erroneous conclusions . Soil properties vary not only from one location to another; the variability may arise from differences in parent materials, vegetation, erosion, drainage, topography, agri-input use, etc. If the variability of a soil is high, greater attention is to be paid in collection of the soil sample. The more heterogeneous the soils, the more intense the sampling must be in order to attain accuracy.

There are two major causes of obtaining non-representative samples: (i) contamination and, (ii) inadequate sampling of a field with variable nutrient status. Contamination usually arises from carelessness, e.g. use of containers for sampling, which were previously used for fertilizers, herbicides or detergent containing N, P or K, and which were not to be used for storage of soil samples. Normal precautions will reduce the danger of contamination. The problem of non-representative samples arising from inadequate sampling is a major problem. Overcoming this problem requires a thorough understanding of the nature and sources of variability in the field. This understanding allows to decide on the number and location of sampling sites for a correct estimate of the population.

## Sampling Methods : Sampling Plans

Four different methods or plans are available for soil sampling:

i. judgement sampling,

ii. simple random sampling,

iii. stratified random sampling,

iv. systematic sampling.

A sampling plan designates, which parts (specific locations) of the land is to be included in the sample. Some plans are more representative than others and some can be executed at lower cost than others. The best design is one, that gives high precision (low error) at relatively low cost.

### *Judgement sampling*

In this plan, the researcher or extension officer selects some part of the land which, in his/her judgement, is typical or representative of the field characteristics. For example, the sample collector may base his/her selection on differences in soil colour, and judge that some particular shades of colour are typical of soils of the site. Also he/she may base it on plant growth pattern. The sample collector or researcher has some knowledge about the soils of the field and uses this knowledge to obtain what he/she considers a representative sample. Any confidence in the results from the samples will have to rest on faith in his/her judgement, which may be good or poor.

Since, sample unit selections have different (varying) but unknown probabilities, sample selections are biased. For example, in an attempt to include as many extremes as possible in the sample, the sample collector may commit the error of over-sampling. On the other hand, the sample collector may exclude all extreme cases, thereby end up with a non-representative sample.

Unless the researcher or extension officer is very skillful and experienced in selecting "typical" sites, this method has low precision. The selection of typical sites is difficult and time consuming, and the sampling method is inaccurate for large sample sites. However, if the sites are small and no estimate of accuracy is needed, this method is satisfactory. Generally, judgement sampling is used, where soil or cropping differences are noticeable and where the focus of the survey is only a particular area of the field.

### *Simple random composite sampling*

This sampling method is more precise and less subject to the bias of the sample collector, than the judgement sampling method. Sampling units are taken by selecting each sampling spot separately, randomly and independently of any spot previously taken. Thus, each unit spot on the field has an equal chance of being selected as the sampling spot. The most common sample collection method, is to take a random sample from the field, without any reference to topography or other field features. This method works well in fields, that are uniform in soil type, production and management history. Although the method is very precise, it is relatively cumbersome and time consuming. For this reason, the method is not used in routine soil testing programmes. It is more adaptable to research work, for sampling experimental plots.

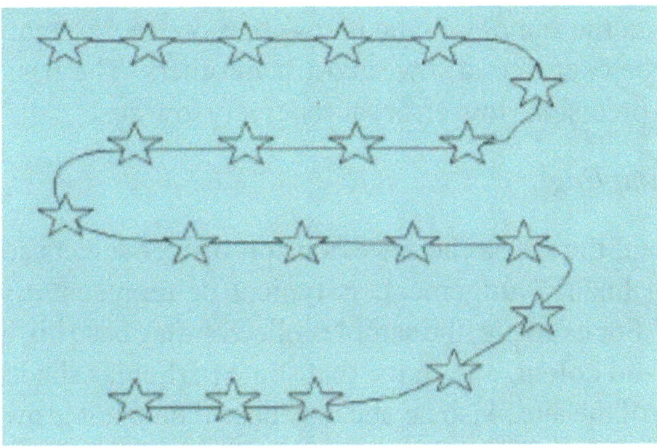

One composite soil sample made up of 20 sub samples

### *Stratified random sampling*

In this method, the field is divided into separate units (strata) based on observable differences in the field. The stratification may be based on soil types or series, land forms (slopes versus plain), soil drainage, crop growth pattern, differences in soil management history, erosion impact, etc. Each of the unit is then sampled separately by random sampling.

This method is usually adopted in routine soil testing programmes. Researchers or scientists also prefer this method in experimental plots because it allows the worker to (i) make a statement about each sub-population, i.e. stratum or sampling unit, separately, and (ii) increase the accuracy of estimates over the entire field (population).

Stratified random sampling is selected for its accuracy (low error) in the sense that, stratification itself is done to eliminate some of the variation from the sampling error. Since, the strata are sampled separately, the differences among strata means are eliminated from the error, and only the within-stratum variation contributes to the sampling error. Although stratification increases precision (and the more the stratification, the greater the increase in precision), excessive stratification for this purpose must be avoided, especially because of increased cost of sampling and analysis. Also, it gets to a point where further stratification adds a very insignificant value to sampling precision. In any case, since at least two units (composites) must be sampled from each stratum to allow an estimation of the sampling error of a stratum, it becomes necessary to keep the number of strata as small as is just enough to allow satisfactory estimates of error.

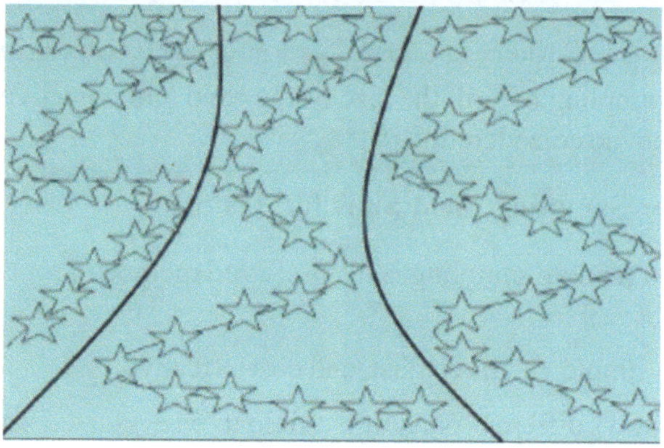

Three composite soil samples, each made up of 20 sub samples

### *Systematic sampling*

This method is very simple to use, and it ensures better coverage of the population (field) than the three other methods. The basic characteristic of this method is that, the selected sampling spots (units) are at regular intervals away from each other, either in one or in two dimensions, thereby forming a grid. The first sampling spot is selected at random, and the subsequent spots selected at uniform intervals.

Some workers have found the method to have about the same precision as simple random sampling and less precision than stratified random sampling. It is known, however, that under certain conditions, the system is inefficient; for example, when there is a fertility gradient on the field or when there is a drainage or slope problem on the field.

Twenty composite soil samples, each made up of 5 sub samples

Before sampling of soil, the size of the field and objectives of sampling should be taken ino consideration.

## Procedure for Sampling Field Soils for Testing

Stratified random sampling is the preferred method in routine soil testing because:

i. it is simple, relatively fast, and cost-effective

ii. it allows different parts of the field to be evaluated separately

iii. it minimizes measurement error due to soil variability

iv. it has relatively good precision

The following procedure is to be adopted

### *Preliminary planning*

a. Sketch a map of the field to be sampled.

b. Outline on this map areas of the field, which are observed to be different from other areas. These differences might be in soil types, degree of erosion, slope, drainage, crop growth performance or past management.

c. Obtain at **least two** composite samples from each sampling unit (stratum). Each composite sample should be from an area that is as uniform as possible. The number of composite samples depends on the size of each sampling unit (strata). The number as a fraction of the total number of samples, is proportional to the size of the stratum as a fraction of the field size.

### *Sampling technique*

a. Take a minimum of 25 and maximum of 40 cores per two-hectare field (depends on observed soil variability).

b. Take at least two composite samples per sampling unit (stratum).

c. Take a minimum of five cores per composite sample (the number depends on size of land represented by the composite sample, as well as the size of the sample container). Enough cores should be taken to represent the whole sampling area adequately.

d. If an area within a sampling unit (stratum) is quite obviously different from the rest of the stratum, but is too small to be of importance, do not sample it (i.e. ignore it). If it is large enough, take a composite sample from the area, but do not mix cores from this with cores from other areas in the stratum.

e. Avoid sampling a fertilizer band if the position of the band is apparent.

f. If possible sample when a crop, preferably a grain crop like maize, rice or wheat, is growing on the field. This enables variability in the field to be identified quite easily.

g. Record the position of each composite sample on the map.

h. Each core should be taken at the same soil depth, so that core samples would be of the same soil volume. Select soil depth according to the purpose of sampling. For arable crops, sample at a depth of 0 to 15 cm. For perennial or tree crops, take deeper samples since tree crops often grow deep into the soil.

### *Sampling tools*

Depending on the purpose and precision required, the following tools may be needed for taking soil samples:

1. a soil auger – it may be a tube, post-hole or screw-type auger or even a spade for taking sample
2. a clean bucket or a tray or a clean cloth – for mixing the soil and sub sampling
3. cloth bags of a specific size
4. a copying pencil for markings, and tags for tying cloth bags
5. soil sample information sheet.

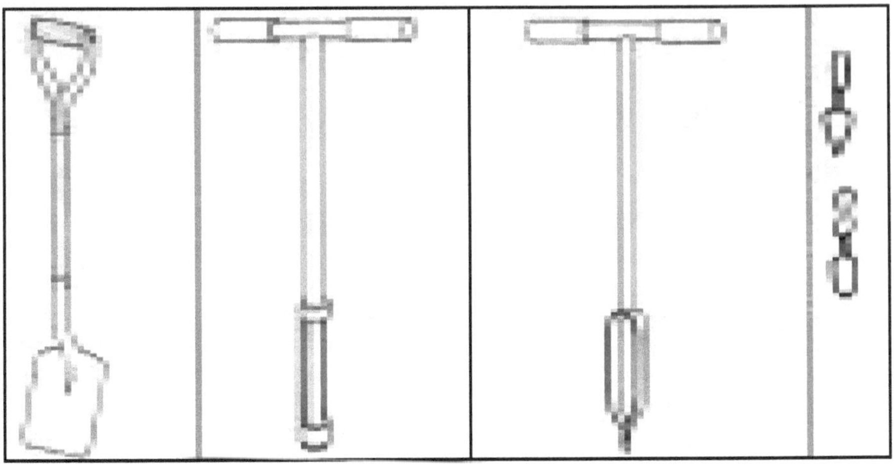

Sampling tools

### *Sampling pattern*

The sampling pattern for each composite sample should be random. The core samples are taken randomly in a zig-zag manner, such that all points in the plot have an equal chance of being selected as a sampling point. Cores should not be taken close to the field boundaries, because, these areas are frequently disturbed and different from the main field. Take soil samples at about the same time of the year, e.g. just before land preparation or before planting. This will allow for comparison of soil test data over periods of time, as there seems to be seasonal variation in soil tests for some nutrients such as P and K.

If a shovel or spade is used instead of a sampling tube, scrape away surface litter to obtain a uniformly thick slice of soil from the surface to the plough depth from each spot. Make a V-shaped cut with a spade to remove a 1–2cm slice of soil. Collect the sample on the blade of the spade and put it in a clean bucket. In this way, collect samples from all the spots marked for one sampling unit. Put all core samples taken from one soil area together in a clean bucket, as a composite sample. Mix the sample thoroughly with your hand. Pour the soil from the bucket onto a piece of clean paper or cloth, and mix it thoroughly. Spread the soil evenly and divide it into quarters. Reject two opposite quarters and mix the rest of the soil again. Repeat the process until left with about 0.5 kg of the soil. Collect it and put in a clean cloth bag. Mark each bag clearly in order to identify the sample. The bag used for sampling must always be clean and free from any contamination. If the same bag is to be used a second time, turn it inside out and remove the soil particles. Write the details of the sample on the information sheet. Put a copy of this information sheet in the bag. Tie the mouth of the bag carefully. Pour the soil sample into a clean plastic bag and tie it securely; label each bag properly for identification later on.

Other related Information should be obtained, if possible, from the farmer regarding: Crops to be grown in the succeeding 3-5 years (if it is a continuous cultivation system), previous crops, previous application of fertilizer and lime, irrigation/credit facilities or personal fund available to the farmer.

***Precautions***

a. Do not sample unusual areas, such as unevenly fertilized areas, marshy areas, old paths, old channels, old bunds, areas near trees, sites of previous compost piles, and other unrepresentative sites.

b. For a soft and moist soil, the tube auger or spade is considered satisfactory.

c. For harder soil, a screw auger may be more convenient.

d. Where crops have been planted in rows, collect samples from the middle of the rows, in order to avoid the area where fertilizer has been band placed.

e. Avoid any type of contamination at all stages. Soil samples should never be stored with fertilizer materials and detergents. Contamination is likely, when the soil samples are spread out to dry in the vicinity of stored fertilizers or on floor where fertilizers were stored previously.

f. Before putting soil samples in bags, they should be examined for cleanliness as well as for strength.

g. The information sheet should be filled in clearly with a copying pencil.

***Sampling salt-affected soils***

Salt-affected soils (including calcareous saline/non-saline soils) may be sampled in two ways. Surface samples should be taken in the same way as for soil fertility analysis. These samples are used to determine the gypsum requirement of the soil. For reclamation purpose, it is necessary to know also the characteristics of the soil profile. Therefore, such soils are sampled down to a depth of 1 m. The samples may be removed from one to two spots per 0.4 ha, where the soil is uniformly salt-affected. If patches are conspicuous, then all large patches should be sampled separately. Soil is sampled separately for soil depths (about 0.5 kg from each depth) of 0–15, 15–30, 30–60 and 60–100 cm. If a stony layer is encountered during sampling, such a layer should be sampled separately and its depth noted. This is very important and must not be ignored. Soil samples can be removed by a spade, or if the auger is used, then care should be taken to note the depth of "concretion" (stones) or the impermeable layer (hardpan). If the soil shows evidence of profile development or distinct stratification, samples should be taken by horizon. If a pit is dug and horizons are absent, then mark the vertical side of the pit at depths of 15, 30, 60 and 100 cm from the surface and collect about 0.5 kg soil from each layer, cutting uniform slices of soil separately. In addition to the above sampling, one surface soil sample should be taken as in the case of normal soil sampling for fertilizer recommendation. Pack the samples and label the bags in the same way as for normal soil sampling, giving additional information about the depth of the sample. The sheet accompanying the sample must include information on:

- nature of the soil
- hardness and permeability of the soil
- cause and source of salinity (where known)
- relief
- seasonal rainfall
- irrigation and frequency of waterlogging
- water table
- soil management history
- crop species and conditions of plant cover
- depth of the hardpan or concretion.

As the salt concentration may vary greatly with vertical or horizontal distance and with moisture and time, it is necessary to keep an account of the time of irrigation and of the amount of irrigation or rain received prior to sampling.

### *Soil sampling for Orchards*

There is often wide variations among soils of an orchard due to organic manure use, fertiliser use, irrigation, differences in elevations etc. With random sampling methods, spatial variability in soil properties pose problems in sampling and interpretation of soil analysis data. For this,grid sampling is used,which provides better information about soil fertility.

For fruit trees planted in a block (mango, litchi, citrus, papaya etc.) 3-4 cores from 0-30cm depth and 75-100 cm distance from the tree trunk are collected. Thus, about 15-20 cores will make up a single soil sample.These are mixed well and then a representative sample is collected.

### *GIS based soil sampling*

It is recommended (Maji, 2014) that soil tesing labs should use soil samples collected from farmers fields to generate the data input for fertility mapping at cadastral level (1:4000 scale). For this,the following steps are essential:

* Use a GPS-based system for soil sample collection to form a geo-referenced baseline for mapping. Then, a laptop can be used in the field for co-lateral data collection and creation of a database.
* The soil analysis database is to be interfaced with the spatial database (the geo-referenced cadastral map) to generate cadastral level soil fertility maps with farmers holding details and other field details.
* The maps need to be placed on an accessible platform to allow users to obtain information on fertility status of respective area/location. This is possible through the networking of Agricultural Research Centres, KVKs, ATMAs and the STLs.

### *Depth, Time & Frequency of Soil Sampling*

For cultivated crops, soil samples are collected from 0-15 or, 0-20cm depth (depth of tillage layer). In lawns and pastures,lime and nutrients are broadcast on the surface. In such cases, soil samples are collected at 5-10cm depth. In low rainfall region, pre-plant soil samples (60-90cm depth) need to be collected.

Normally, soil samples are collected just before planting or early in the crop growth cycle. But, for routine tests,soil samples can be collected anytime before/after rains during non-crop period.

Soil test recommendations are generally valid for 3 years. But, in light textured soils, the frequency of soil tesing may be less. For pH change monitoring, 3-year period is sufficient. Closer monitoring may be required for nutrients, such as P or Zn to make changes in their recommendation for use in crops.

### *Dispatch of soil samples to the laboratory*

Before sending soil samples to the testing laboratory, it is necessary to ensure that proper identification marks are present on the sample bags and labels placed in the bags. It is essential to use a copying pencil/ball pen and not ink because ink can smudge and become illegible. The best system is to obtain soil sampling bags from the soil testing laboratory with most of the information printed or stencilled on them in indelible ink. Compare the number and details on the bag with the dispatch list. The serial numbers of different places should be distinguished by putting the identification mark specific for each centre. This may be in letters, e.g. one for the district, another for the block/panchayat, and a third for the village. Pack the samples properly. Wooden boxes are most suitable for long transport. Sample bags should be packed only in clean bags never used for fertilizer or detergent packing. Farmers may bring soil samples directly to the laboratory. However, most samples are sent to the laboratories through field extension staff. An organized assembly, processing and dispatch system is required in order to ensure prompt delivery of samples to the laboratory.

### *Handling in the laboratory*

As soon as the samples arrive at the soil testing laboratory, these should be checked against the accompanying information list. If the laboratory personnel have collected the samples themselves, then adequate field notes should have been kept. All unidentifiable samples should be discarded. Information regarding samples should be recorded in a register, and each sample should be given a laboratory number, in addition to the sample number, to help distinguish where more than one source of samples is involved.

### *Drying of samples*

Samples received in the laboratory may be moist. These should be dried in wooden or enamelled trays. Care should be taken to maintain the identity of each sample at all stages of preparation. During drying, the trays can be numbered or a plastic tag could be attached. The samples are allowed to dry in the air. Alternatively, the trays may be placed in racks in a hot-air cabinet, whose temperature should not exceed 35 °C and whose relative humidity should be 30–60 per cent. Oven drying a soil can cause profound changes in the sample. This step is not recommended as a preparatory procedure despite its convenience. Drying has a negligible effect on total N content, but the nitrate content in the soil, changes with time and temperature. Drying at a high temperature affects the microbial population. With excessive drying, soil K may be released or fixed depending on the original level of exchangeable K. Exchangeable K will increase if its original level was less than 1 me/100 g soil (1 cmol/kg) and vice versa, but the effect depends on the nature of clay minerals in the soil. In general, excessive drying, such as oven drying of the soil, affects the availability of most of the nutrients present in the sample and should be avoided. Only air drying is recommended.

For analysis of nitrate, nitrite and ammonium in soil, the samples should not be dried and determinations must be carried out on samples brought straight from the field. However, the results are expressed on an oven-dry basis by estimating separately the moisture content in the samples.

### *Post-drying care*

After drying, the samples are taken to the preparation room. Air-dried samples are ground with a wooden pestle and mortar, so that the soil aggregate is crushed, but the soil particles do not break down. Samples of heavy (clay) soils may have to be ground with an end-runner grinding mill fitted with a pestle of hard wood and rubber lining to the mortar. Pebbles, concretions and stones should not be broken during grinding. After grinding, the soil is sieved through a 2mm sieve. The practice of passing only a portion of the ground sample through the sieve and discarding the remainder is erroneous. This introduces a positive bias in the sample, as the rejected part may include soil elements with differential fertility. Therefore, the entire sample should be passed through the sieve except for concretions and pebbles of more than 2 mm. The coarse portion on the sieve should be returned to the mortar for further grinding. Repeat sieving and grinding until all aggregate particles are fine enough to pass the sieve and only pebbles, organic residues and concretions remain. If the soil is to be analysed for trace elements, containers made of copper, zinc and brass must be avoided during grinding and handling. Sieves of different sizes can be obtained made of stainless steel. Aluminium or plastic sieves are useful alternative for general purposes. After the sample has passed through the sieve, it must be mixed again thoroughly. The soil samples should be stored in cardboard boxes in wooden drawers. These boxes should be numbered and arranged in rows in the wooden drawers, which are in turn fitted in a cabinet in the soil sample room.

### *Soil analysis*

The following estimations are generally carried out in a service oriented soil testing laboratory:

- soil texture
- soil structure
- cation exchange capacity (CEC)
- soil moisture
- water holding capacity
- pH
- lime requirement
- electrical conductivity
- gypsum requirement
- lime requirement
- organic C
- total N
- mineralizable N
- inorganic N

- available P
- available K
- available S
- calcium
- magnesium
- micronutrients – available Zn, Cu, Fe, Mn, B and Mo.

Soil sampling is an useful farm management tool, but it is important to keep in mind the soil test results are only as accurate as the sampling technique and the records kept on each sample. A good soil fertility program requires regular soil sampling and accurate record keeping and it may also require fertilizer, manure or inputs and plant tissue testing for deriving meaningful inferences. A sampling program that includes the preparation of a farm map each year outlining the location of each sample and the crop management practices that were associated with each field, is recommended. Once you have chosen a soil testing lab, it is a good idea to stick with that lab, because each individual laboratory has its own "soil testing philosophy" for the determination of soil nutrient levels and also for the interpretations and recommendations that come from the test results.

## Reference

Maji, A.K.(2014) GIS in soil fertility evaluation for making fertiliser recommendations. In Soil Testing for Balanced Fertilisation (ed.HLS Tandon). Fetiliser Development and Consultation Organisation, New Delhi, India. pp 27-48.

❑❑❑

# B. Instrumental Methods

## Basic Chemistry in Soil Chemical Analysis

**Molarity:** One molar (M) solution contains one mole or one molecular weight in grams of a substance in each litre of the solution.

**Normality:** The normality of a solution is the number of gram equivalents of the solute per litre of the solution. It is usually designated by letter N. Semi-normal, penti-normal, desi-normal, centi-normal and milli-normal solutions are often required, these are written (shortly) as 0.5N, 0.2N, 0.1N, 0.01N and 0.001N, respectively. The definition of normal solution utilizes the term 'equivalent weight'. This quantity varies with the type of reaction, and hence it is difficult to give a clear definition of equivalent weight which will cover all reactions. It often happens that the same compound possess different equivalent weights in different chemical reactions.

**Buffer solutions:** Solutions containing a weak acid and its salt or weak base and its salt (e.g. $CH_3COOH + CH_3COONa$) and ($NH_4OH + NH_4Cl$) possess the characteristic property to resist changes in pH, when some acid or base is added in them. Such solutions are referred to as buffer solutions. The important properties of a buffer solution are i) It has a definite pH value ii) Its pH value does not alter on keeping it for a long time and iii) Its pH value is only slightly altered when strong base or strong acid is added. It may be noted that because of the above property, readily prepared buffer solutions of known pH are used to check the accuracy of pH meters.

**Titration:** It is a process of determining the volume of a substance required to just complete the reaction with a known amount of other substance. The solution of known strength used in the titration is called titrant. The substance to be determined in the solution is called titrate. The completion of the reaction is judged with the help of appropriate indicator.

**Indicator:** A substance, which indicates the end point on completion of the reaction is called an indicator. Most commonly used indicators in volumetric analysis are **i) Internal indicators,** like methyl red, methyl orange, phenolphthalein and diphenylamine, which are added in the solution, where reaction occurs, are called internal indicators. On completion of the reaction of titrant on titrate, a colour change takes place due to the presence of indicator, which also helps in knowing that the titration is complete. The internal indicators used in acid - alkali neutralization solutions are methyl orange, phenolphthalein and bromothymol blue. The indicator used in precipitation reactions like titration of neutral solution of NaCl (or chloride ion) with silver nitrate ($AgNO_3$) solution is $K_2CrO_4$. Redox indicators are also examples of internal indicators, which possess different colours in the oxidized and reduced forms. For example, Diphenylamine has blue violet colour under oxidation state and colourless in reduced condition. Ferroin gives blue colour under oxidation state and red colour under reduced condition. **ii) External indicators,** are indicators used outside the titration mixture. Potassium ferricyanide is used as an external indicator in the titration of potassium dichromate and ferrous sulphate in acid medium. In this titration, few drops of indicator are placed on a white porcelain tile. A glass rod dipped in the solution being titrated is taken out and brought in contact with the drops of indicator placed on white tile. In the beginning deep blue colour is noticed, which turns greenish on completion of titration. **iii) Self indicator,** after completion, the reaction leaves its own colour due to its slight excess presence. In $KMnO_4$ titration with ferrous sulphate, the addition of $KMnO_4$ starts reacting with $FeSO_4$, which is colourless. On completion of titration, slight excess presence of KMnO4 gives pink colour to the solution, which acts as a self indicator and points to the completion of the titration.

**Standard solution:** The solution of accurately known strength (or concentration) is called a standard solution. It contains a definite number of gram equivalent or gram mole per litre of solution. If it contains 1 gram equivalent weight of a substance/compound, it is 1N solution. If it contains 2 gram equivalent weights of the compound, it is 2N. All volumetric methods depend upon standard solutions, which contain known amounts (exact) of the reagents in unit volume of the solution. A solution is prepared, having approximately the desired concentration. This solution is then standardized by titrating it with another substance, which can be obtained in highly purified form. Thus, potassium permanganate solution can be standardized against sodium oxalate, which can be obtained in a high degree of purity, since it is easily dried and is non-hygroscopic. Such substance, whose weight and purity is stable, is called as 'Primary Standard'. A primary standard must have the following characteristics:

- It must be obtainable in a pure form or in a state of known purity.
- It must react in one way only under the condition of titration and there must be no side reactions.
- It must be non-hygroscopic. Salt hydrates are generally not suitable as primary standards.
- Normally, it should have a large equivalent weight so as to reduce the error in weighing.

- An acid or a base should preferably be strong, that is, they should have a high dissociation constant for being used as standards.

Primary standard solution is one, which can be prepared directly by weighing and with which other solutions of approximate strength can be titrated and standardized. Generally, Potassium hydrogen phthalate and Sodium carbonate is used as an acid and base, respectively in soil testing laboratories. Secondary standard solutions are those, which are prepared by dissolving a little more than the gram equivalent weight of the substance per litre of the solution and then their exact standardization is done with primary standard solution. Standard solutions of all the reagents required in a laboratory must be prepared and kept ready before taking up any analysis. However, their strength should be periodically checked or fresh reagents be prepared before analysis.

### *Titration*

In all titrations involving acidimetry and alkalimetry, standard solutions are required. These may be prepared either from standard substances by direct weighing or by standardizing a solution of approximate normality of materials by titrating against a prepared standard. The method of preparation of standard hydrochloric acid solution is given below:

Concentration of conc. HCl is approximately 11N.

Therefore, to prepare a standard solution (0.1N), it is diluted roughly one hundred times. Take 10 ml of acid and make approximately to 1 litre by dilution with distilled water. Titrate this acid against 0.1N $Na_2CO_3$ (Primary standard) using methyl orange as indicator. Colour changes from pink to yellow, when acid is neutralized. Suppose 10 ml of acid and 1.5ml of $Na_2CO_3$ are consumed in the titration. $V_1N_1=V_2N_2$; $V_1 = 10$ ml; $N_1$=unknown; $V_2 = 1.5$ ml; $N_2 = 0.1$ then $N_1$ calculated as 0.015.

Similarly, normality of sulphuric acid can be worked out. $H_2SO_4$ needs to be diluted about 360 times to get approximately 0.1N, because it has a normality of approximately 35 (calculated based on specific gravity, purity and molecular weight mentioned on the bottle). Then, titrate against standard $Na_2CO_3$ to find out exact normality of $H_2SO_4$.

## *Basic calculations*

**Percent concentration, specific gravity, Normality and amount needed for making 1N solution of some commonly used reagents in STL**

| Reagent | Empirical Formula | Formula weight | Purity (%) | Normality | Molarity | Density | ml required to prepare 1 litre of 1 N solution |
|---|---|---|---|---|---|---|---|
| Acetic acid, glacial | $CH_3000H$ | 60.06 | 99.7 | 17.4 | 17.4 | 1.05 | 6.9 |
| Acetic acid | $CH_3000H$ | 60.06 | 80.0 | 14.3 | 14.3 | 1.07 | 9.5 |
| Hydrochloric acid | HC1 | 36.46 | 37.0 | 12.1 | 12.1 | 1.19 | 23.2 |
| Hydrofluoric acid | HF | 20.01 | 48.0 | 27.6 | 27.6 | 1.15 | 18.9 |
| Nitric acid | $HN0_3$ | 63.01 | 90.0 | 21.1 | 21.1 | 1.48 | *47.0* |
| Nitric acid | $HNO_3$ | 63.01 | 70.0 | 15.7 | 15.7 | 1.41 | 42.2 |
| Nitric acid | $HNO_3$ | 63.01 | 65.0 | 14.3 | 14.3 | 1.39 | 40.7 |
| Perchloric acid | $HC1O_4$ | 100.47 | 70.0 | 11.6 | 11.6 | 1.67 | 58.2 |
| Perchloric acid | $HC10_4$ | 100.47 | 60.0 | 9.2 | 92 | 1.54 | 50.8 |
| Phosphoric acid | $H_3PO_4$ | 98.00 | 85.0 | 44.0 | 14.7 | 1.69 | 59.2 |
| Sulfuric acid | $H_2SO_4$ | 98.08 | 95.0 | 35.6 | 17.8 | 1.84 | 66.2 |
| Ammonium hydroxide | $NH_4OH$ | 35.05 | 57.6 | 14.8 | 14.8 | 0.90 | 25.6 |

**Some important conversion factors**

| | | |
|---|---|---|
| N x 1.286 = $NH_4^+$ | N x 4.43 = $NO_3^-$ | Organic carbon x 1.724 = Organic matter |
| P x 2.29 = $P_2O_5$ | K x 1.20 = $K_2O$ | mm per opening of sieve = 16/ mesh per inch |

ppm (parts per million) = micro gm per ml = mg per litre = mg per kg=%age x 10000.

**Procedure for preparing 1000 ppm standard solutions**

| Element | Procedure for making 1 litre standard solution |
|---|---|
| Ca | Dissolve 2.4973 gm of $CaCO_3$ in 25 ml of 1 N HCl, and make the volume 1 litre |
| P | Dissolve 4.39 gm of AR grade Potassium dihyrogen phosphate in 100 ml distilled water and to it add 5 ml of conc. Sulphuric acid and makeup the volume 1 litre. |
| K | Dissolve 1.9067 gm of AR grade KCl in 100 ml distilled water and make the volume 1 litre |
| S | Dissolve 5.434 gm of AR grade potassium sulphate in 100 ml of distilled water and make the volume 1 litre. |
| Fe | Dissolve 1.00 gm of Fe metal (99.99% pure) in 100 ml of 3.5 N $H_2SO_4$ and make the volume 1 litre. |
| Cu | Dissolve 1.00 gm of Cu metal (99.99% pure) in 50 ml of 1:1 $HNO_3$ and make the volume 1 litre. |
| Mn | Dissolve 1.00 gm of Mn metal (99.99% pure) in 50 ml of 6 N HCl and make the volume 1 litre. |
| Zn | Dissolve 1.00 gm of Zn metal (99.99% pure) in 50 ml of 1:1 HCl and make the volume 1 litre. |

## Principles of Analytical Instruments, Their Calibration and Application

### *I. pH Meter*

A pH meter is basically a high impedance amplifier, that accurately measures the minute electrode voltages and displays the results directly. A glass electrode in contact with H+ ions of the solution acquires an electric potential, which depends on the concentration of $H^+$ ions. This is measured potentiometrically against some reference electrode, which is usually a calomel electrode. The potential difference between glass electrode and calomel electrode is expressed as pH. pH describes the degree of acidity or alkalinity of a solution. It is measured on a scale of 0 to 14. The pH scale of the device should be calibrated by at least two buffer solutions. Usually one of the buffers used for calibration has pH 7.00 and the second is selected depending on the range, where the measurements are to be taken at 9.2 for basic solutions and 4.0 for acidic solutions.

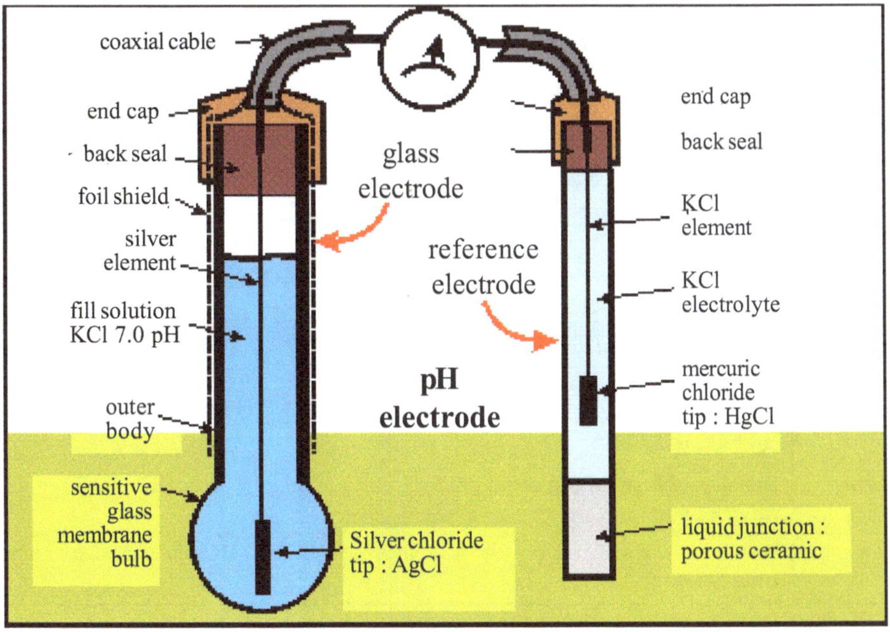


The electrode consists of mercury in contact with a solution of potassium chloride saturated with mercurous chloride. It maintains a constant potential, at a given temperature. In commercial models, a paste of mercury and $Hg_2Cl_2$ is contained in an inner tube connected to the KCl solution in an outer jacket. The lead wire is connected to the $Hg_2Cl_2$ paste through a column of mercury. The outer tube ends in a fine capillary to provide a salt bridge through the test solution to the glass electrode. pH meter with single (actually combined) electrode is also available as in case of digital type instrument.

The advantage of glass electrode is that, it can be used in any solution not being affected by organic compounds or oxidising and reducing agents. A small quantity of solution is sufficient for determination of the pH. Special glass membranes are required, when pH of the solution is very high (pH > 10). Such special electrodes are also commercially available, where sodium of the glass is replaced by lithium. The apparent e.m.f./pH slope will be 59.15 mV per pH unit at 25°C using the equation pH = pHs–(E–Es)/0.000198T, where pHs and Es are the values in the standard state and T is the absolute temperature in K. It utilizes the Nernst equation.

The direct reading type of instrument, although possibly less accurate than potentiometric is also used exclusively in modern soil laboratories. The e.m.f. of the glass electrode-calomel electrode cell is applied across a resistance, and the resulting current after amplification is passed through an ammeter causing deflection of the pointer across a scale marked in pH units. These instruments are available to operate on mains A.C. current. In most pH meters, temperature control knob is provided to adjust at temperature of the test solution. Liquid junction potential is the most important source of error, when using the glass electrode, calomel electrode system. When two solutions of different strength or composition come into contact, the more concentrated solution will diffuse into the more dilute one. If the ions of the diffusing solution move at different speed, the dilute solution will assume an electric charge with respect to the concentrated solution corresponding to that of the faster moving ion. For example, if the diffusing anions move more quickly than the cations, they will cause the dilute solution to become negative with respect to the concentrated solution. The resulting difference in potential across the interface of the solutions is called the **'liquid junction potential'** (Ej) and adds to or subtracts from the electric potential. Such a potential is likely to arise at the liquid junction between a soil suspension, and the salt bridge of the calomel electrode. The presence of colloids or suspensions has a marked effect on, liquid junction potentials and hence this error may be more important, in soil pH measurements than when using pure solutions. Attempts have been made to allow for liquid junction potentials by calculation. The calculation involves knowledge of activity coefficients and even for true solutions has proved to be of little use and would be quite impossible to derive for soil suspensions. One procedure to minimize the liquid junction potential is to use saturated potassium chloride solution as the salt bridge. It is the relative mobilities of the oppositely charged ions at the interface that decide the potential gradient and thus, it is desirable to equate these mobilities' as far as possible. Potassium chloride is used as potassium ions and chloride ions have about the same mobility, and if the concentration of the salt is greater than that of other electrolytes present, it will be responsible for transferring almost all the current across the liquid junction.

*Standardization of pH meter*

1. Switch on the instrument and allow it to warm up for 10 minutes.
2. Keep the pH selector switch on zero position.
3. Set the temperature compensation control to the solution temperature.
4. Remove the electrode from distilled water, wipe it dry with filter paper and dip in standard buffer solution of known pH (4.0, 7.0 and 9.2).
5. Press the stdby / read button to 'READ' position.
6. Adjust the calibration knob to the required pH of the buffer solution.
7. Press the stdby / read button to 'STDBY' position.
8. Remove the electrode from the buffer and wash with distilled water and wipe it dry with filter paper.
9. Continue the calibration in a similar way for all the buffer solutions.

*Precautions*

1. Wash the electrode every time after each sample or buffer solution is tested.
2. Electrode should not touch the sides/ base of the container.

*General tips*

1. The electrode should always remain immersed in the water.
2. Allow the instrument to stabilize for a period of at least 30 minute before use.
3. Change the water in the beaker daily.
4. If there is strong decline or fluctuation in the reading, fill the electrode with saturated KCl solution.

**Interpretation of soil pH**

| S.No. | pH range | Rating |
|---|---|---|
| 1. | <4.5 | Very strongly acidic |
| 2. | 4.5 – 5.5 | Strongly acidic |
| 3. | 5.5 – 6.5 | Slightly acidic |
| 4. | 6.5 – 7.5 | Neutral |
| 5. | 7.5 – 8.5 | Slightly alkaline |
| 6. | 8.5 – 10.0 | Strongly alkaline |
| 7. | >10.0 | Very strongly alkaline |

### *II. Conductivity Meter*

Electrical conductivity is the measure of the ability of the solution to carry electric current by the migration of ions under the influence of an electric field and called as salt bridge or solubridge or conductivity meter. Like metallic conductors, solutions also obey ohm's law. A conductivity meter measures the ionic conductivity (or conversely, the resistance) of a liquid. The device usually consists of a probe, which usually has two platinum electrode plates parallel to each other and separated by some small distance. The specific conductance (L or K) or conductivity of a solution is always obtained by measuring the resistance (R) of the solution taken in a suitable container of known dimensions called conductivity cell, the cell constant of which has been determined by calibration with a solution of accurately known conductivity e.g. a standard KCl solution. The instrument used for electrical conductivity measurement is known as conductivity bridge. A typical system consists of an alternating current (A.C.) Wheatstone bridge, a primary element of conductivity cell and a null balance indicator (as in 'solubridge') or an electronic eye as in the conductivity meter.

*Reporting the levels*

Conductivity is customarily reported in micro-mhos per centimeter (mmhos/cm). In the international system of units the reciprocal of the ohm is the simens (S) and the conductivity is reported as milli-siemens or desi-siemens per meter (mS/m or dS/m).

**Conductivity of standard solutions** : Conductivity is measured against standard solution of KCl and following concentrations are used for the calibration.

| Solution type | Mass of KCl / litre | Conductivity at 25 $^0$C | Temp. coefficient at 25 $^0$C |
|---|---|---|---|
| 1.0 M | 71.1352 g | 111.3 μS/cm | 1.89 % |
| 0.1 M | 7.4191 g | 12.85 μS/cm | 1.90 % |
| 0.01 M | 0.7452 g | 1408 μS/cm | 1.94 % |
| 0.001 M | 0.0753 g | 146.1 μS/cm | 2.04 % |

The above table also can be used to calculate the conductivity at any other ambient temperature.

Example: Calculation of conductivity of 0.01 M KCl at 30 $^0$C

$$C\ 30\ ^0C = C\ 25\ ^0C + C\ 25\ ^0C \times 1.94/100 \times (30\ ^0C - 25\ ^0C)$$
$$= 1408\ \mu S + 1408 \times 1.94/100 \times 5$$
$$= 1544.5\ \mu S/cm$$

### *III. UV-Visible Spectrophotometer*

Spectrophotometer is concerned with the determination of the concentration of a substance by measurement of relative absorption of light with respect to a known concentration of the substance. It enables one to measure absorbance (or, transmittance) at various wavelengths. A spectro-photometer may also be regarded as a refined filter photoelectric photometer, which permits the use of continuously variable and more nearly monochromatic bands of light. The essential parts of a spectrophotometer are *(i)* a source of radiant energy, *(ii)* a monochromator (filter, prism or diffraction grating) *i.e.* a device for isolating monochromatic light *i.e.* light of a single frequency or more precisely expressed narrow bands of radiant energy from the light source *(iii)* glass or silica cells for the solvent and for the solution under test and *(iv)* a device to receive or measure the beam or beams of radiant energy passing through the solvent or solution in terms of electricity generated. The instrument used in Ultraviolet-visible spectroscopy is called as ultraviolet-visible **spectrophotometer.** To obtain absorption information, a sample is placed in the spectrophotometer and ultraviolet or visible light at a certain wavelength, or range of wavelengths, is transmitted through the sample. The spectrophotometer measures how much of the light is absorbed by the sample. A brief overview of a spectrophotometer is given below.

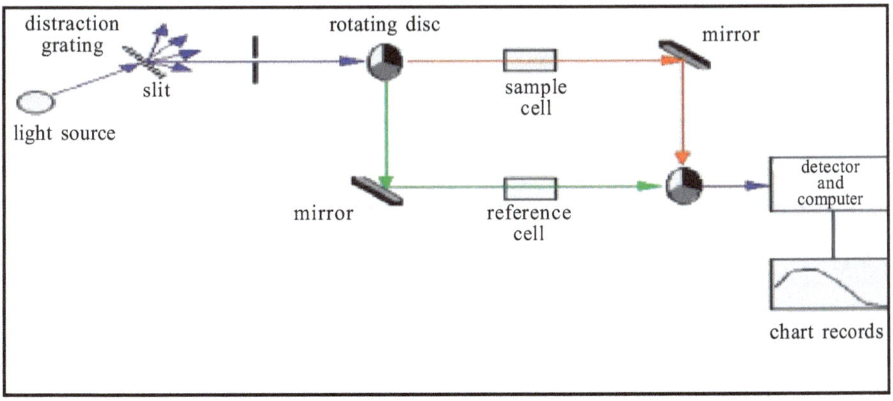


UV/VIS spectroscopy is routinely used in the quantitative determination of solutions of transition metals and highly conjugated organic compounds. It is possible to do so because transition metals are often coloured, because of the possibility of d-d electronic transitions within the metal atoms. Organic molecules, especially those with a high degree of conjugation also absorb light in the UV or visible regions of the electromagnetic spectrum.

**Selection of wavelength range**

| Wavelength range (nm) | Transmitted colour | Complementary hue |
|---|---|---|
| 380 – 435 | Violet | Yellowish green |
| 435 – 480 | Blue | Yellow |
| 480 – 490 | Greenish blue | Orange |
| 490 – 500 | Bluish green | Red |
| 500 – 560 | Green | Purple |
| 560 – 580 | Yellowish green | Violet |
| 580—595 | Yellow | Blue |
| 595 – 650 | Orange | Greenish blue |
| 650 – 780 | Red | Bluish green |

The functioning of this instrument is relatively straightforward. A beam of light from a visible and/or UV light source (coloured red) is separated into its component wavelengths by a prism or diffraction grating. Each monochromatic (single wavelength) beam in turn is split into two equal intensity beams by a half-mirrored device. One beam, the sample beam (colored magenta), passes through a small transparent container (cuvette) containing a solution, the compound being studied in a transparent solvent. The other beam, the reference (colored blue), passes through an identical cuvette containing only the solvent. The intensities of these light beams are then measured by electronic detectors and compared. The intensity of the reference beam, which should have suffered little or no light absorption, is defined as $I_0$. The intensity of the sample beam is defined as I. Over a short period of time, the spectrometer automatically scans all the component wavelengths in the manner described. The ultraviolet (UV) region scanned is normally from 200 to 400 nm, and the visible portion from 400 to 800 nm. Most of the spectrophotometers display absorbance and the commonly observed range is from 0 (100% transmittance) to 2 (1% transmittance). The wavelength of maximum absorbance is a characteristic value, designated as $l_{max}$. Different compounds may have very different absorption maxima and absorbance. Intensely absorbing compounds must be examined in dilute solution, so that significant light energy is received by the detector, and this requires the use of completely transparent (non-absorbing) solvents. Spectrophotometry or colorimetry is based on Beer and Lambert's law, which states that when monochromatic light passes through a transparent medium, intensity of the emitted light decreases exponentially as the concentration/thickness of the absorbing medium increases arithmetically. The decrease in the intensity of light is directly proportional to the concentration of the solution if the thickness of the absorbing medium is kept constant.

$$\text{Transmittance } T = I/Io$$

Where, Io is the intensity or energy per $cm^2$ per second before passing through substance. I is the intensity of energy after passing through the medium

$$\text{Absorbance (A)} = \log 1/T$$
$$= \log Io/I$$
$$= \log Io - \log I$$
$$= \log 100 - \log T$$
$$A = 2 - \log T$$

Spectrophotometer

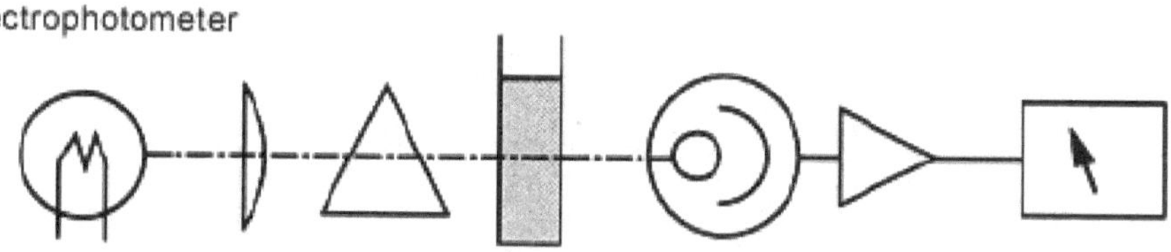


A calibration curve or standard curve is usually prepared by using series of standard solutions of known concentrations and the curve is used to determine the concentration of unknown samples. The usual method of use of spectrophotometer requires the construction of standard curve for the constituent being determined. Suitable quantities (concentration) of the constituent are taken and treated in the same way as the sample solution for the development of colour and the measurement of the transmittance (or absorbance) was carried out at the particular wavelength and plotted by taking absorbance in Y-axis and concentration in X-axis. This equation of this line by first finding the slope ($m = (y_2 - y_1)/(x_2 - x_1)$) and changing that into $y = mx + c$, for this experiment the slope and intercept of that line provides a relationship between absorbance and concentration that determined by Beer's Law. From the absorbance obtained for the samples their corresponding concentration is obtained.

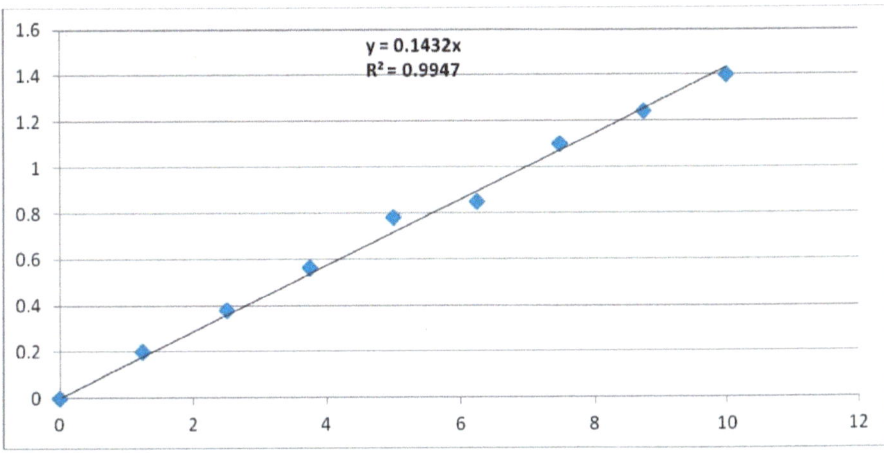


Concentration (ppm)

The above graph, where concentration is plotted against absorbance then a straight line (Beer's Law) is fit to the data that is obtained and the resulting equation is used to convert absorbance of the unknown sample into concentration. Colour developed can also be matched depending upon the concentration.

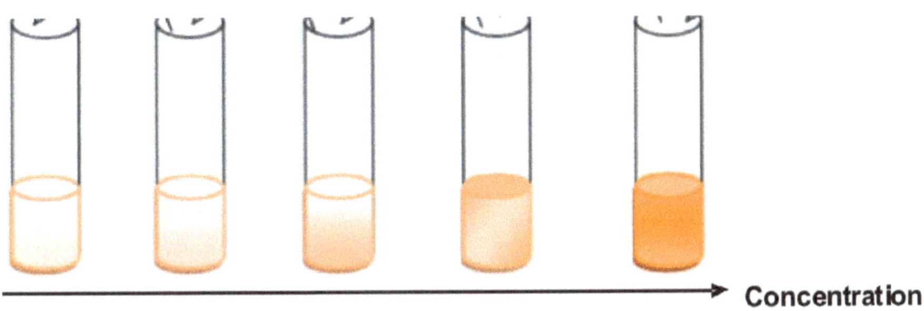


***IV. Flame Photometer***

Atomic spectroscopy is thought to be the oldest instrumental method for the determination of elements. These techniques were introduced in the mid of

19th Century, during which Bunsen and Kirchhoff showed that the radiation emitted from the flames depends on the characteristic element present in the flame. The potential of atomic spectroscopy in both the qualitative as well as quantitative analysis, were then well established. Developments in the field of instrumentation led to varied application of atomic spectroscopy. Atomic spectroscopy is an unavoidable tool in the field of analytical chemistry. It is divided into three types, which are absorption, emission, and luminescence spectroscopy. The different branches of Atomic Absorption Spectroscopy are (1) Flame photometry or flame atomic emission spectrometry, in which the species are examined in the form of atoms (2) Atomic absorption spectrophotometry, (AAS), (3) Inductively coupled plasma-atomic emission spectrometry (ICP-AES).

Flame photometers use atomic emission for the routine detection of metal salts, principally sodium (Na), potassium (K), lithium (Li) and calcium (Ca). A flame photometer is an instrument used for measuring the spectral intensity of metals present in the metallic salt.

Flame photometry, now more properly called flame atomic emission spectrometry is a relatively old instrumental analysis method. Its origin dates back to Bunsen's flame colour tests for the qualitative identification of select metallic elements. A flame photometer is an instrument, in which the intensity of the filtered radiation from the flame is measured with a photoelectric detector. The filter interposed between the flame and the detector, transmits only a strong line of the element.

Analytical flame photometry is based on the measurement of the intensity of the characteristic line emission of the element to be determined. When a solution of a salt is sprayed into a flame (liquefied petroleum gas), the salt gets separated into its component atoms because of the high temperature. The energy provided by the flame excites the atoms to higher energy levels. Actually the orbital electrons are shifted to higher planes from their normal orientation. When the electrons return back to ground state or unexcited state, they emit their characteristic radiation. Since the excitation can be to different levels, light (electromagnetic radiation) of several wavelengths can be emitted. However, the intensity of the wavelength corresponding to the most probable transition will be the highest. For each element, such characteristic lines have already been well identified. The comparison of emission intensities of unknown samples to either that of standard solutions (plotting calibration curve), or to those of an internal standard (standard addition method), helps in the quantitative analysis of the metal in the sample solution. Flame emissions of the alkali and alkaline earth metals in terms of the emission wave length and the characteristic colour produced by each element is shown in table 1.

**Table 1 :** Characteristic colour and wavelength of elements analysed by flame photometer

| Name of the element | Emitted wavelength range (nm) | Observed colour (flame) |
|---|---|---|
| Potassium (K) | 766 | Violet |
| Lithium (Li) | 670 | Red |
| Calcium (Ca) | 622 | Orange |
| Sodium (Na) | 589 | Yellow |

*Parts of a flame photometer*

**1. Source of flame :** A burner that provides flame and can be maintained in a constant form and at a constant temperature.

**2. Nebuliser and mixing chamber** : Helps to transport the homogeneous solution of the substance into the flame at a steady rate. The solution of the substance to be analysed is first aspirated into the burner, which is then dispersed into the flame as fine spray particles.

**3. Optical system (optical filter)** : The optical system comprises of three parts: convex mirror, lens and filter. The convex mirror helps to transmit light emitted from the atoms and focus the emissions to the lens. The convex lens helps to focus the light on a point called slit. The reflections from the mirror pass through the slit and reach the filters. This isolates the wavelength to be measured from that of any other extraneous emissions. Hence, it acts as interference type color filters.

**4. Photo detector** : It detects the emitted light and measures the intensity of radiation emitted by the flame. That is, the emitted radiation is converted to an electrical signal with the help of photo detector. The produced electrical signals are directly proportional to the intensity of light.

*A brief overview of the processes in a flame photometer*

**Desolvation**: The metal particles in the flame are dehydrated by the flame and hence the solvent is evaporated.

**Vapourisation**: The metal particles in the sample are dehydrated. This also leads to the evaporation of the solvent.

**Atomization**: Reduction of metal ions in the solvent to metal atoms by the flame heat.

**Excitation**: The electrostatic force of attraction between the electrons and nucleus of the atom helps them to absorb a particular amount of energy. The atoms then jump to the exited energy state.

**Emission process**: Since the higher energy state is unstable, the atoms jump back to the stable low energy statAnd, the most effective BCAs studied to date appear to antagonize pathogens using multiple mechanisms. For instance, pseudomonads known to produce the antibiotic 2,4-diacetylphloroglucinol (DAPG) may also induce host defenses. Additionally, DAPG-producers can aggressively colonize roots, a trait that might further contribute to their ability to suppress pathogen activity in the rhizosphere of wheat through competition for organic nutrients e with the emission of energy in the form of radiation of characteristic wavelength, which is measured by the photo detector.

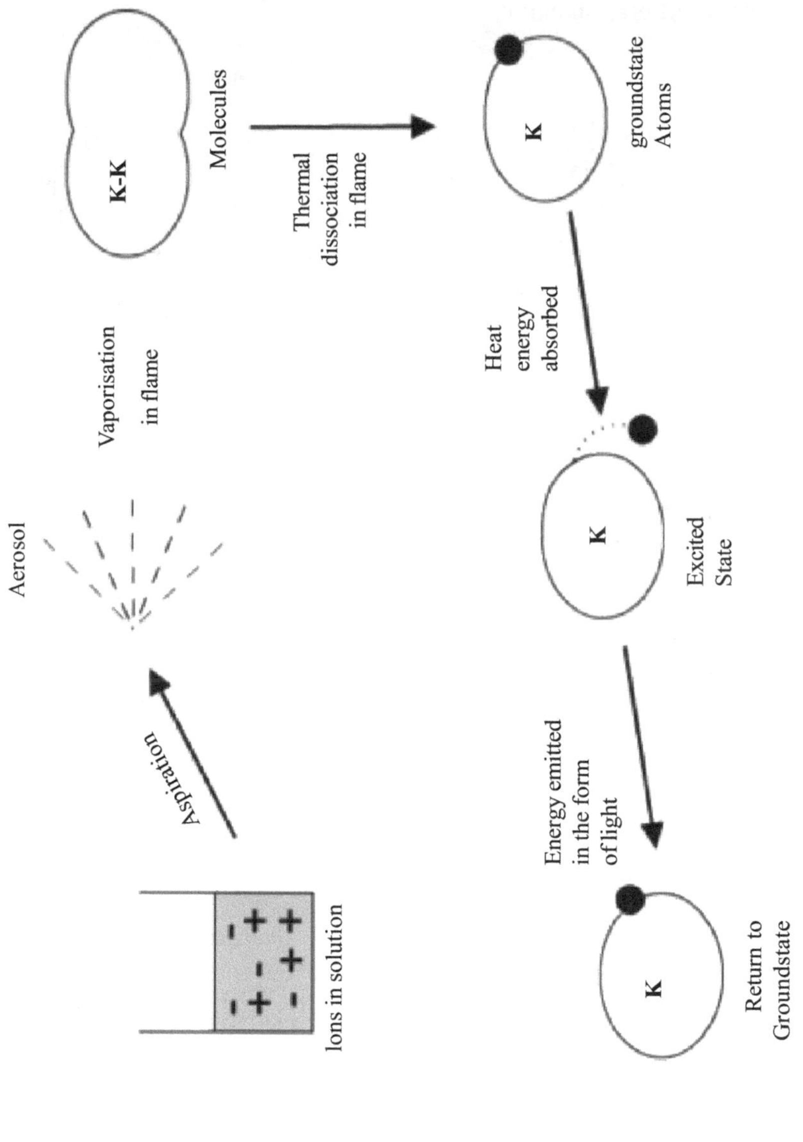


Overview of the processes in a Flame photometer

Each individual atom emits one quantum of radiation, therefore, the intensity of radiation emitting from the flame will be proportional to the number of atoms in the flame, that is, to the concentration of the particular element in the flame. The comparison of emission intensities of unknown samples to either that of standard solutions (plotting calibration curve), or to those of an internal standard (standard addition method), helps in the quantitative analysis of the metal in the sample solution. A series of standard solutions are prepared and the intensity of emission determined for each concentration after zero setting of blank and hundred setting of the maximum concentration. The intensity of emissions from the test solutions is measured simultaneously and the concentration of the element is read from the calibration curve.

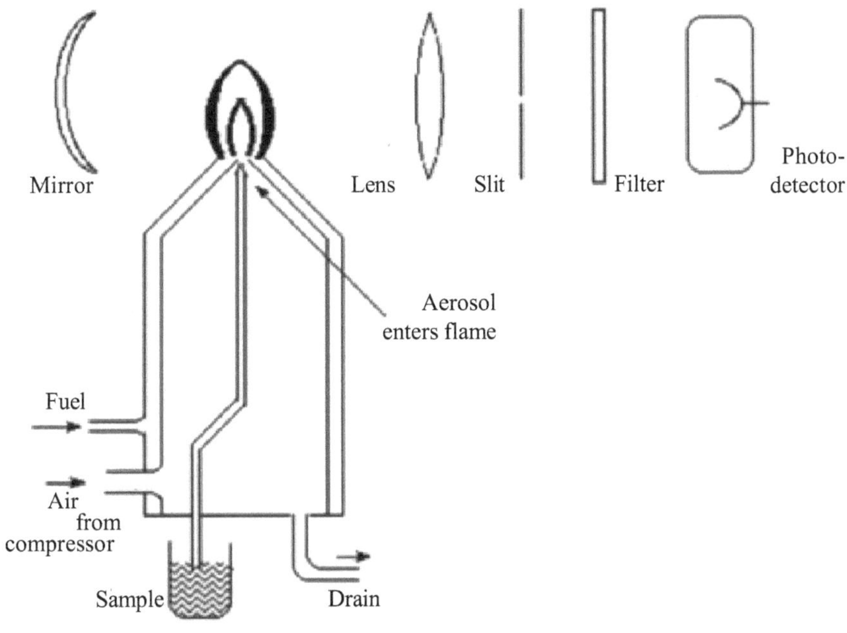


Schematic diagram of a flame photometer

*Advantages*

1. Simple quantitative analytical test based on the flame analysis.
2. Inexpensive.
3. The determination of elements, such as, alkali and alkaline earth metals is performed easily with most reliable and convenient methods.
4. Quite quick, convenient, selective and sensitive to even parts per million (ppm) range.

*Disadvantages*

1. A standard solution with known molarities is required for determining the concentration of the ions, which will correspond to the emission spectra.
2. It is difficult to obtain the accurate results of ions with higher concentration.
3. The information about the molecular structure of the compound present in the sample solution cannot be determined.
4. The elements such as carbon, hydrogen and halides cannot be detected due to its non radiating nature.

***V. Atomic Absorption Spectrophotometer***

Atomic absorption spectrophotometer (AAS) is an analytical equipment based on atomic absorption spectro-photometry and is used to measure metals in the sample. When a sample is aspirated into the instrument, it is subjected to a heavy thermal environment and as a result, 'the ground state' atoms absorb light energy of a specific wavelength and enter into the excited sate. As the number of atoms in the light path increases, the amount of light absorbed increase in a predictable way. By measuring the amount of light absorbed, a quantitative determination of the amount of the element present can be made.

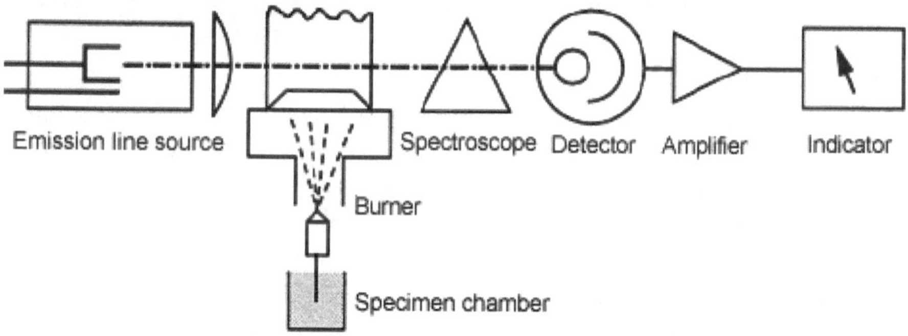


In AAS, atoms excited in an oxy-acetylene flame absorb light at wavelengths characteristic of each element. Light absorption is proportional to the concentration of atoms in a light path. The most commonly used radiation source is a hollow cathode lamp, consisting of a tungsten anode and a cylindrical cathode sealed in a glass tube filled with neon or argon gas. AAS has five basic components which are (1) A light source (hollow cathode lamp); (2) A sample cell (absorption cell); (3) Monochromator; (4) Detector; and (5) Output unit.

Absorption in the flame is by **vapour phase atoms**, giving absorption line spectra. A continuous spectrum light source, even with high quality monochromator cannot achieve sufficiently narrow band pass width for absorption line spectra. Use of special lamps, each **emitting line spectrum** matched to the line spectrum of the analyte atoms in the flame. The type of lamp is a **hollow cathode lamp.** Separate lamps for each analyte element is required, but multi-element lamps are also available.

Though, both single-beam and double-beam instruments are available, double-beam instruments are preferred, because they minimize effects of lamp emission variations, detector sensitivity, and electronic gain. Acetylene ($C_2H_2$) as fuel and air or nitrous oxide ($N_2O$) as oxidant are commonly used. An air-$C_2H_2$ flame (2100-2400°C) is used for elements (such as calcium, magnesium), that do not form refractory compounds and have low ionization potentials; an $N_2O$-$C_2H_2$ flame (2600-2800°C) is used for elements (such as aluminum) forming refractory compounds.

More than 60 elements can be determined by AAS. Besides detection limits, factors such as matrix or interference effects are major influences on the viability of particular analysis. Problems caused by the formation of stable compounds or compounds of low volatility by the element of interest in combination with someone can be overcome by the addition of an excess of a "competing action" or a "releasing agent". Lanthanum (1000 ppm') as chloride can be added in the determination of calcium and magnesium to remove the potential interference due to aluminum and phosphorus. To overcome ionization interference, an excess of an easily ionized metal is added. For example, 1000 ppm Cesium is added to solutions for the determination of potassium and sodium. To eliminate matrix interferences, sample extracts and standard solutions should be in the same solvent.

Instrument settings are normally those recommended by the manufacturer; however, slight adjustments *might* be necessary in *certain* instances. The burner is flushed with water and the "zero absorption" re-established before aspirating each sample. At least three standards are prepared to obtain a working curve. Working *curves* are prepared with each batch of samples.

If the concentration of the element of interest is above the analytical range of the instrument, then the solution must be diluted. An alternative is to place the burner head perpendicular to the light path (or at any other angle that may be required). This increases the upper limit of linearity by a factor of upto 10.

*Makeup three standards*

The first one should be at the top of the linear range. The concentration of the second standard should be approximately 3 times the concentration of the first. The concentration of the third standard should be approximately 6 times the concentration of the first standard.

### *Characteristic Concentration vs. Detection Limit*

Characteristic concentration in atomic absorption (sometimes called "sensitivity") is defined as the concentration of an element (expressed in mg/l) required to produce a signal of 1% absorption (0.0044 absorbance units). As long as measurements are made in the linear working range, characteristic concentration can be determined by reading the absorbance produced by a known concentration of the element, and solving the following equation.

Characteristic Concentration = Conc. of Std. (mg/l) 0.0044

The characteristic concentration check value is the concentration of element (in mg/l), that will produce a signal of approximately 0.2 absorbvance units under optimum conditions at the wavelength listed.

### *VI. Microwave Digestion System*

Microwave digestion system is an advanced and highly sophisticated system for sample preparation, which utilizes the electromagnetic radiation to achieve higher temperature for the reaction and provides high performance, reliable quality and safety, which is required in sample preparation in order to achieve superior analytical results. Its closed vessel technique helps to speed up reactions by allowing higher temperatures, while preventing the loss of volatile analytes. The resulting low reagent consumption, saves time and money and also helps to minimize exposure to corrosive gases and hazardous solvent vapors. The magnetron of the system generates microwaves of 0.3 mm to 1.0 m wavelength having frequency of 100 GHz to 300 MHz, which has strong penetrating power on the matrix of extract the analyte of interest.

This instrument is now widely used in domestic, commercial and scientific segments for a variety of purposes. In the field of research, it is specifically employed in digesting of various abiotic and biotic matrices to extract organic and inorganic chemicals.

### *VII. Auto Kjeldhal Nitrogen Analyzer*

Nitrogen determination has a long history in the area of analytical chemistry. Johan Kjeldahl first introduced the Kjeldahl nitrogen method in 1883. While studying proteins during malt production, he developed a method of determining nitrogen content, that was faster and more accurate than any method available at the time. Since 1883, the Kjeldahl method has gained wide acceptance and is now used for a variety of applications, Kjeldahl nitrogen determinations are performed on food and beverages, meat, feed, grain, waste water, soil and many other samples.

The auto Kjeldhal nitrogen analyzer consists of two separate units, block digester and distillation, titration assembly. In block digester, digestion is performed at 420 $^{0}C$. Further, digested samples are transferred into distillation titration assembly, where librated ammonia is collected in boric acid solution and titrated against standard acid. After the end of process, result can be read out on display.

*The Kjeldahl method may be broken down into three main steps:*

**Digestion** – the decomposition of nitrogen in organic samples utilizing a concentrated sulfuric acid. The end result is an ammonium sulfate solution.

Organic N + $H_2^-SO_4 \rightarrow (NH_4)_2SO_4^- + H_2O + CO_2$ + other sample matrix by products.

**Distillation** – Adding excess base to the acid digestion mixture to convert $(NH_4)_2SO_4$ to $NH_3$, followed by boiling and condensation of the $NH_3$ gas in a receiving solution.

$$(NH_4)_2SO_4 + 2NaOH \rightarrow 2NH_3 + Na_2SO_4 + 2H_2O$$

$$NH_3 + H_3BO_3 \rightarrow NH_4^+ : H_2BO_3^- + H_3BO_3$$

**Titration** – To quantify the amount of ammonia in the receiving solution, the amount of nitrogen in a sample can be calculated from the quantified amount of ammonia ions in the receiving solution.

$$NH_4^+ : H_2BO_3^- + HCl \rightarrow NH_4Cl + H_3BO_3$$

$$N\ (\%) = \frac{(S - B)\ N\ 1.40}{\text{Sample weight (g)}}$$

Where,

S = Volume of acid used against sample.

B = Volume of acid used against blank.

N = Normality of acid.

If it is desired to determine % protein instead of % nitrogen, the calculated % N is multiplied by a factor, the magnitude of the factor depending on the sample matrix. Many protein factors have been developed by AACC and AOAC for use with various types of samples such as, 6.38 for milk and dairy, 5.95 for rice, 5.70 for wheat flour and 6.23 for other grains.

## *VIII. Inductively Coupled Plasma Atomic Emission Spectroscopy (ICP-AES)*

An inductively coupled plasma-atomic emission spectroscopy (ICP-AES) spectrometer consists of an excitation source (RF generator), a sample introduction system (nebulizer, spray chamber, and torch), an optical resolving system (primary slit, diffraction grating, secondary optics, and photomultipliers), and an electronic data capture and storage system (measuring electronics and microcomputer).

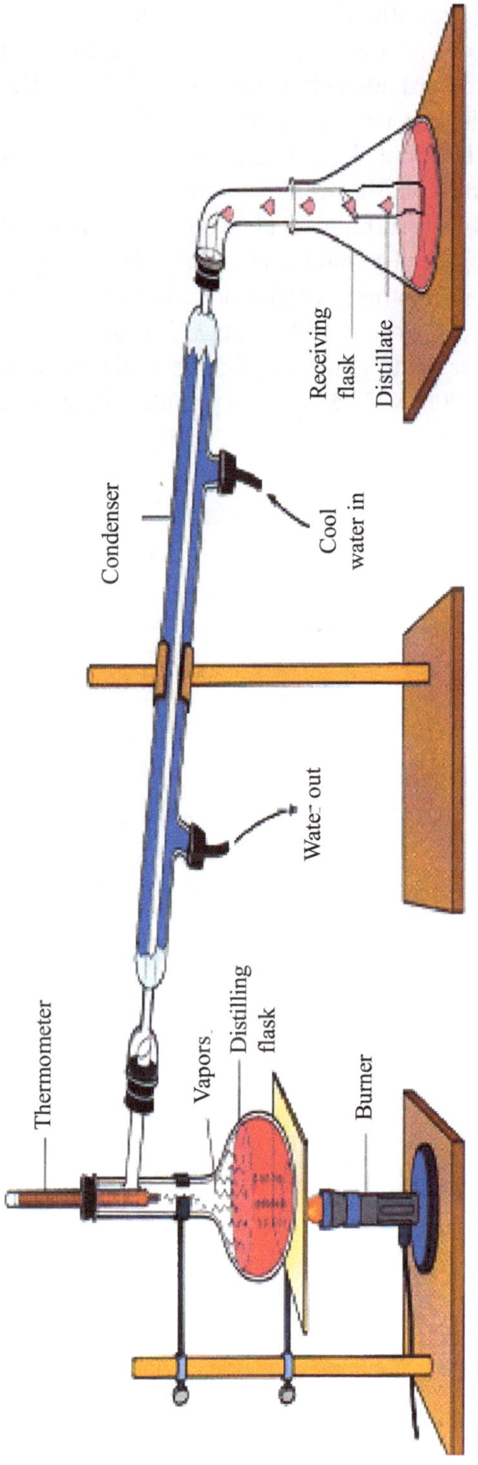


Glass assembled Kjeldahl distillation unit

The ICP is produced by passing initially ionized Ar gas through a quartz torch located inside Cu coil connected to a radio frequency (RF) generator. The RF generator provides up to 3 KW forward power at a frequency of 27.1 MHz. The high frequency currents flowing in the Cu coil generate oscillating magnetic fields, whose lines of force are axially oriented inside the quartz tube and that follow elliptical closed paths outside the coil. Electrons and ions passing through the oscillating electromagnetic field flow at high acceleration rates in closed annular paths inside the quartz tube space. The induced magnetic fields vary in their direction and strength, resulting in electron acceleration on each half cycle. Collision between accelerated electrons or ions and unionized Ar gas further causes ionization. The collisions cause heating, which produces a plasma with temperatures ranging from 6000 to 10000°K. The resultant plasma is contained within the torch by means of argon flow.

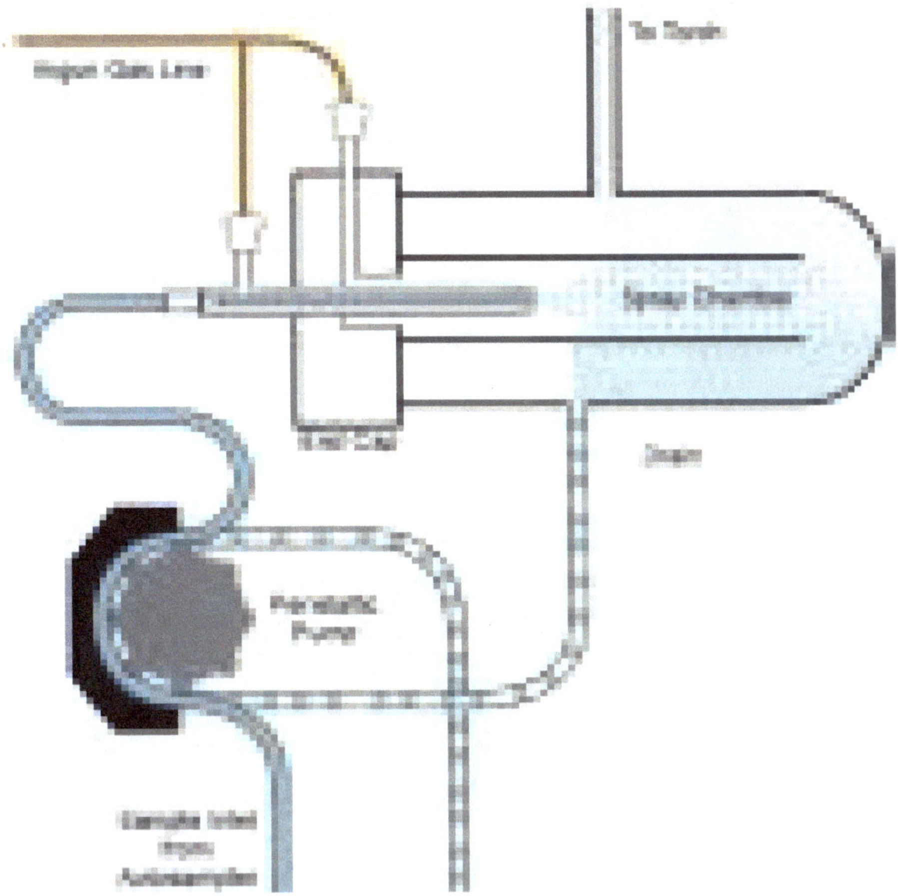

An overview of the processes in ICP-AES

The method of presenting the sample to the plasma is similar to that used in atomic absorption. The liquid sample is aspirated by a nebulizer, that atomizes the sample and presents the aerosol to the torch via the spray chamber. The spray chamber serves to settle out larger droplets, allowing the smaller droplets to enter the torch assembly. In the torch, the sample is injected into the plasma, causing the excited neutral atoms or ions within the sample to emit radiation of specific wavelengths. The emitted energy is observed by means of the light tube. It can be focused on the entrance slit of a spectrometer and illuminates the diffraction grating. Here, the light is separated into its component wavelengths and passes through an exit slit, where it is order-sorted and impinged on a photo multiplier tube. The photo multiplier produces a signal directly proportional to the intensity of the impinging light. This signal is then passed by the measuring electronics to the computer. The data are compared with previously stored data from standards and converted to concentration data. The concentration data are stored on disks and presented to the operator at the input-output device. Some of the advantages of ICP-AES are as follows:

1. Inter- elemental interference is low.
2. Good spectra for many elements under a single set of excitation conditions, allowing multi-element analysis of very small aliquots.
3. Low concentrations of elements (e.g., boron, phosphorus), that tend to form refractory compounds can be determined.
4. Elements over a dynamic range of concentration, four to six orders of magnitude, may be determined.
5. Ionization interference effects are small or non-existent.
6. Analytical stability is long-term.
7. Detection limits are comparable or better than other atomic spectral procedures. Soil extracts and plant digests can be analysed directly without the pre-concentration techniques, often required for flame atomic absorption spectrophotometry.
8. Cost per sample analysed is low.

The following points should be considered in ICP-AES analysis:

1. For any element, a minimum concentration five times of the detection is required to obtain reliable results.
2. To prevent clogging of the nebulizer, soil extracts and plant tissue digests must be free of suspended particles. This is achieved by filtering with 0.45 filter.
3. "Blank" solutions are used as the zero point.

Generally, three to six standards are used for calibration. These working standards should be in the same matrix as the samples. Pure reagents and double distilled water should be used for the preparation of standard stock solutions.

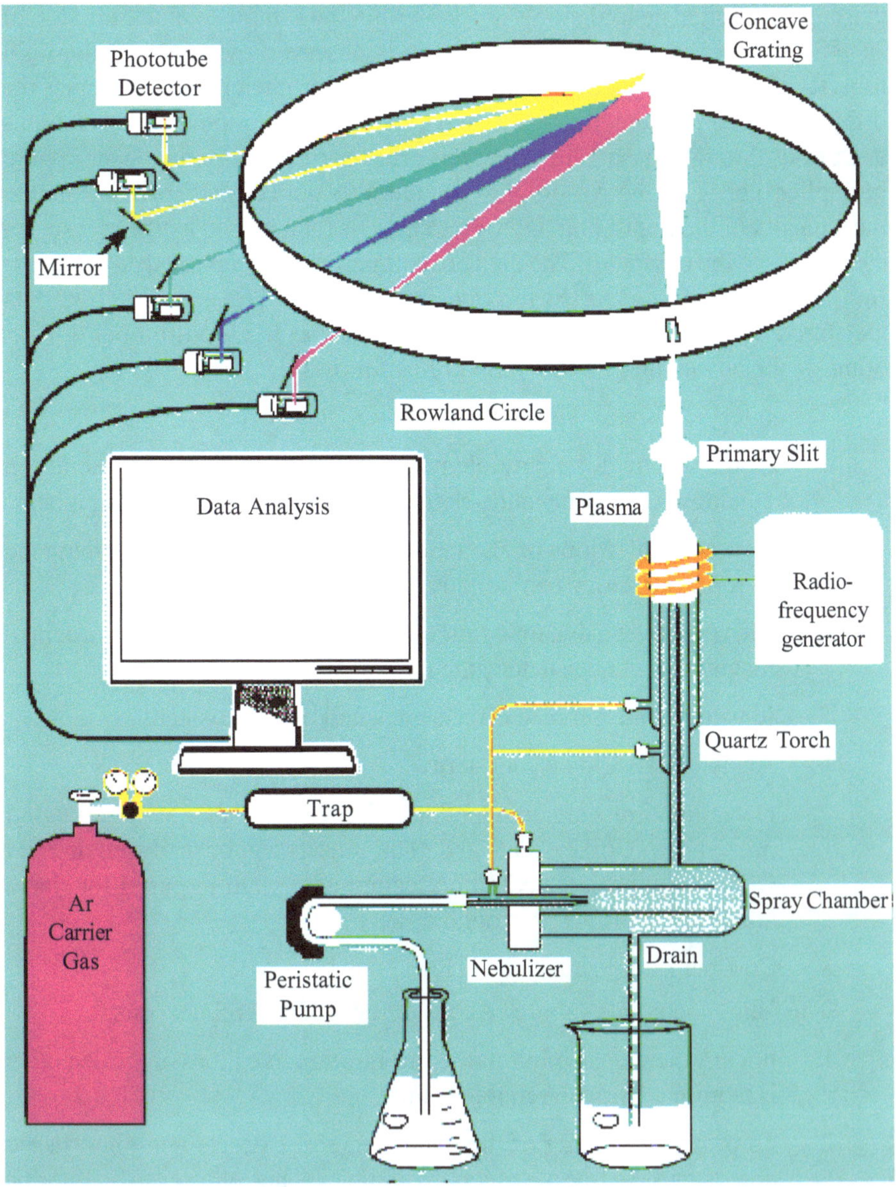


Schematic diagram of ICP-AES

Spectrometer softwares supplied by the manufacturer enables the operator to do many tasks automatically. Automated sample introduction permits the analysis of a large number of samples without operator attention. A wash cycle is chosen, that is sufficient to eliminate carryover from sample to sample.

❑❑❑

# C. Methods of Analysis

Plant available nutrient in soil at a given point of time comprises only a small fraction of the total nutrient. This available fraction in soil is highly dynamic in nature and is in equilibrium with certain other forms. Therefore, a reliable index of soil fertility is determined through soil testing. Among other things, the success of soil testing depends upon the methodology adopted and interpretation of test results. A suitable method is the one, which satisfies following three criteria i.e. (i) It should be fairly rapid so that the test results can be obtained in a reasonably short period (ii) It should give accurate and reproducible results of a given sample with least interference during estimation and, (iii) It should have high predictability i.e. a significant relationship of the test values with crop performance. While efforts have been made all over the world to develop universally acceptable methods, such as a single soil test for assessing nutrient availability in soils of varying characteristics is yet to be evolved. In fact, the soil test methods are situation-specific in nature.

Soil testing has been widely accepted as an essential tool in formulating a sound fertilizer programme. But, many misconceptions prevail, as to what soil tests can or can not do. Some consider soil testing as little more than gimmick and is used to promote fertilizer sales, provide personal prestige or influence administrative or legislative bodies. Whereas, others consider it as an accurate and indispensable tool for the assessment of soil fertility status to form the basis, on which fertilizer needs are determined. The divergent views on this important aspect are often due to absence of a sound soil testing programme. Soil testing programme may be divided into four phases i.e. (i) collection of soil samples, (ii) extracting and determing the available nutrients, (iii) calibrating and interpreting the analytical results and (iv) making fertilizer recommendations. The success of a soil testing programme depends upon the operative precision in all the four stages. Unfortunately, when errors are made at any stage of the programme, the chemical analysis becomes irrelevant. In India, introduction of high yielding varieties, chemical fertilizers and soil testing service were contemporary. The former two gained tremendous popularity, but soil testing lagged behind. One of the reasons for this, is that of the soil testing service in India is, lack of adequate background research. A sound soil testing programme requires an enormous amount of background research. The background research should determine (i) the significant chemical forms of the available nutrients in the soils of the area, (ii) the extractants most suitable for accurately measuring the available form of nutrients, (iii) the relative productive capacity of the soils for various crops, the differential response of the various rates and methods of fertilizer application for different crops, (iv) field sampling techniques and methodology etc. The second most important reason for the poor progress of the soil testing programme is, too much emphasis on the number of soil samples collected and analysed and very little attention to quality of analysis, calibration and interpretation of the soil test values. Another factor which is very important for the success of a soil testing programme is, to consider the socio-economic and other local factors while making the fertilizer recommendations. Thus, following may be considered as pre-requisites for successful soil testing service i.e. (i) effective educational activities, (ii) uniform reporting system, (iii) improved sampling techniques, (iv) improved analytical procedures, (v) better correlation and calibration of data, (vi) consistent interpretation and recommendations of fertilizers and (vii) timely delivery of the test results.

## Pre-processing of Soil Samples for Analysis

### *Labeling*

Label samples for identification. A label of thick paper with identification mark and other details should be put inside the sample bag and another label carrying same details tied/ pasted outside the bag. In case, the sample is wet, the label should be written with pencil or a permanent marker. In addition to the location, field number, name of cultivator and identification mark, relevant information about slope, drainage, irrigation facilities, previous cropping history, rate of fertilizer & manure used, choice of crops etc. must be recorded and sent along with the soil samples.

### *Storage*

Once collected, samples for nutrient analyses may be bulked to give a composite sample. Generally, bulking should be done only when the samples come from a relatively uniform area, or what is thought to be a relatively uniform area. Most determinations are made on air-dried samples. In some cases, however, $NH_4$-N, $NO_3$-N, pH, electrical conductivity, and some other properties are determined on moist samples (field condition) immediately after arrival at the laboratory. Drying some soils, particularly organic horizons, can cause irreversible changes in some properties. Stored samples must be tightly closed. In some instances, it might be necessary to air dry part of the sample and to maintain the other part in the field moist state. Problems associated with obtaining a representative sample of moist soils can be reduced by blending moist samples prior to sub- sampling.

Soil samples should be air-dried soon after collection to prevent microbial changes. Soils are air-dried at 20-25°C and with relative humidity of 20-60%, the term "air-dried" refers to soil conditioned to ambient temperature and humidity. Large lumps of moist soil are broken by hand and spread on paper in a room free of fumes, dust, etc. If large clods are not broken, they will take an unduly long time to dry and may also be difficult to grind. When dry, the soil is rolled gently with a wooden roller. Coarse concretions, stones and pieces of macro-organic matter (roots, leaves, and other vegetative material) are picked out.

### *Grinding*

Grinding is essential to homogenize the soil and reduce sub- sampling error as well as to increase the specific surface. After air-drying, the soil is ground to pass through a 2-mm sieve. The grinder consists of three cylinders into which the samples and metal pestles are placed. The cylinders are rotated horizontally by electrically driven rollers. As the cylinders rotate, the sample is ground by the pestle and falls through the mesh of the cylinder walls into a tray below. Remaining gravel (weathered and non-weathered rock fragments) and organic residue (e.g., fibrous material from roots) are removed. These materials are weighed and their percentage in the total sample is determined. Approximately 500 g of homogenized sub-sample fine earth (less than 2 mm soil) is obtained by the quartering method. The process is repeated as many times as necessary. The soil sample is stored preferably in clean cloth or polythene bags with a proper label for identification.

Nearly all determinations are carried out on the fine earth fraction (less than 2 mm). If less than 1 g of sample is required for a particular analysis, then the 2-mm fraction might not be sufficiently representative, then 25-50g of 2 mm fraction is further ground with pestle and mortar to pass through 0.5 or 0.2 mm sieve.

The composition of the grinding and sieving apparatus is important, particularly if trace elements are to be determined. The soil is ground in an agate or porcelain mortar with a pestle, then passed through a nylon 2-mm sieve (or smaller if required). Iron, copper, and brass sieves are avoided. Treatment with a metallic grinder can also result in serious contamination for some analyses (e.g., iron can interfere with organic carbon determination).

Grinding is performed using clean, dry equipment. The grinder must be thoroughly cleaned between samples to avoid carryover. When grinding with a mortar and pestle, the complete sub-sample must be ground to pass the sieve and none is discarded.

## 1. Determination of Soil pH by Glass Electrode Method

### *Principle*

pH may be defined as the negative logarithm of $H^+$ concentration expressed in g/I. It is conventionally defined by the following equation:

pH = - Log $[H^+]$ in which activity of $H^+$ is expressed as g ions / litre. The hydrogen ion concentration can be calculated from the following equation :

When pH = 7 then pH = - log $[H^+]$ Or, 7 = - log $[H^+]$

Or, log $[H^+] = 7 \times 1$

Or, log $[H^+] = - 7 \times \log 10$

Or, log $[H^+] = \log 10^{-7}$

$H^+ = 10^{-7}$

= 0.0000001 g/l.

The $H^+$ concentration increases 10 times per one unit decrease of pH value and similarly decreases 10 times per one unit increase of pH value.

The effective concentration of hydrogen ions includes all sources such as, those arising by dissociation of soluble acids and those dissociation from soil particles. The most important chemical property of soil i.e. pH is due to the activity of active hydrogen ions of the soil solution. The variations of soil pH from soil to soil are due to the physical, chemical and biological characteristics of the soils. For soil samples, pH is normally determined in soil suspension using water or 0.01M $CaCl_2$ or KCl solution as solvent. The ratio of soil to solvent in soil suspension varies from 1:1.5, but 1:2-2.5 is considered ideal. A soil may be acidic, neutral or alkaline. A pH of 6.5 – 7.5 is considered to be the pH range in which most nutrients are available to plants.

The glass electrode generally used for pH measurement employs a glass membrane. The glass electrode is widely used hydrogen ion responsive electrode. When electrode is immersed into a soil suspension, an electrical potential is developed, which produces an impulse and accordingly the galvanometric pointer gives direct reading of pH value of the soil suspension.

**Reagents:** Standard buffer solution: prepare buffer solutions of pH 4.0, 7.0 and 9.2 by dissolving commercially available buffer tablets of above pH values in a 100 ml volumetric flask with freshly prepared distilled water.

### *Apparatus*

pH meter with glass electrode with calomel reference electrode.

### *Procedure*

1. Weigh 10 g of air dry soil sieved through 2 mm sieve and transfer to a clean 100 ml beaker.
2. Add 25 ml of distilled water..
3. Stir the contents intermittently using glass rod and allow it to stand for exactly half an hour.

4. Wash the electrode carefully with a jet type wash bottle containing distilled water and wipe it dry with a piece of filter paper/tissue paper.
5. Calibrate the pH meter with buffer solutions.
6. Stir the soil suspension again just before taking the reading.
7. Immerse the electrode into the beaker containing soil water suspension and press the STDBY / READ button to 'READ' position.
8. Record the meter reading and note the pH of the soil suspension.

Rating of soil with reference to pH

| Rating | pH |
|---|---|
| Acidic | <6.5 |
| Neutral | 6.5-7.5 |
| Alkaline | >7.5 |

***Precautions***

1) The electrode should not be allowed to remain in the test solution or suspension longer than necessary.
2) Immediately after testing, the electrode must be washed off with a strong stream of distilled water from a wash bottle.
3) For storage, after cleaning, the electrode should be suspended in distilled water and the system should be protected from evaporation. Drying out the electrode should be avoided.
4) The glass and reference electrode should be thoroughly washed with distilled water after each measurement.
5) Before sample reading, pH must be adjusted with buffer solution.

***Note:*** *pH in 1:2.5 soil-water suspension is generally used in a soil testing laboratory for recommendation of amendments and fertilisers. In saline & alkali soils, to mask the variability in salt content of soil pH is measured using 0.01 M* $CaCl_2$ *solution.*

**2. Determination of Electrical Conductivity (EC)**

***Principle***

All fertile soils have small amount of soluble salts in them. Rivers, lakes and wells contain varying amounts of dissolved salts. Runoff waters from soil carry soluble salts as well as suspended solids. The total soluble salts can be determined as a electrical conductivity (EC) of the soil suspension/paste / irrigation water using a conductivity meter as 25 $^0$C. The EC is directly proportional to the amount of salts (ions) and expressed in $dSm^{-1}$ (same as, milli mhos $cm^{-1}$). Electrical conductivity is normally determined in the same soil suspension as described for soil pH. However, for deriving accurate information related to salts in problem soils, EC is also measured in saturated paste and the soil solution derived from the paste under suction using a buchner funnel.

**Reagents: 0.01 N KCl solutions:** Weigh 0.7456 g dry KCl and dissolve in distilled water and make the volume to 1 litre. The solution gives EC value of $1411 \times 10^{-3}$ or 1.41 $dSm^{-1}$ at 25 $^0$C (if cell constant is 1.0).

***Apparatus***

EC-TDS analyzer

***Procedure***

1. Use the same soil-water suspension used for pH estimation.
2. Calibrate the instrument using standard solutions of KCl.
3. Immerse the conductivity cell into the solution and take the readout.

   (Generally, readout is in micro/millimhos $cm^{-1}$)

***Calculations***

Salt concentration is me/L = EC ($dSm^{-1}$) at 25 $^0$C × 640.

When the EC of soil is less 1.0-3.0 $dSm^{-1}$, it is considered as satisfactory, >3.0 $dSm^{-1}$ as injurious to crop plants.

## 3. Determination of Organic Carbon

The importance of organic carbon estimation lies in the fact, that it gives an indication of the organic matter content of the soil, which is an index of soil N availability. The organic carbon content of soil, is reported directly as percentage of C or calculated as organic matter by multiplication with a factor of 1.724 assuming that soil organic matter contains on an average 58 percent carbon, so that 100/58 = 1.724 (Van Bemmelen factor). There is also a close relationship between organic carbon and nitrogen content of the soil (C: N ratio). Two methods **i.** Rapid titration method (Walkley and Black, 1934), and **ii.** Colorimetric method (Datta *et al.*, 1962) are generally followed for determination of organic carbon. In both the methods, organic matter is oxidized with Chromic acid ($K_2Cr_2O_7$ + $H_2SO_4$).

***i. Rapid titration method (Walkley and Black, 1934)***

In this method, sulphuric acid is used to provide the heat of dilution about $125^0C$ temperature required and $CO_2$ evolved. In this method, the excess $K_2Cr_2O_7$ or unconsumed $K_2Cr_2O_7$ is determined by back titration with Ferrous ammonium sulphate (redox titration).An empirical recovery factor of 0.77 was used by the original authors. The reaction is as follows:

$$K_2Cr_2O_7 + 4H_2SO_4 = K_2SO_4 + Cr_2(SO_4)_3 + 4H_2O + 3,O$$

$$2\ K_2Cr_2O_7 + 8\ H_2SO_4 + 3\ C = 2\ K_2SO_4 + 2\ Cr_2(SO_4)_3 + 3\ CO_2 + 8\ H_2O$$

$$K_2Cr_2O_7 + 6\ FeSO_4 + 7\ H_2SO_4 = K_2SO_4 + 2\ Cr_2(SO_4)_3 + 3\ Fe_2(SO_4)_3 + 7\ H_2O$$

Thus, $2\ K_2Cr_2O_7 \equiv 36$ g of C

Or, 588 g $K_2Cr_2O_7 \equiv 36$ g of C

Or, 12 litres of 1N $K_2Cr_2O_7 \equiv 36$ g of C

Or, 1 ml of 1N $K_2Cr_2O_7 \equiv 0.003$ g of C

*Reagents*

1. Normal potassium dichromate solution (Dissolve 49.04 g of AR grade potassium dichromate in 500 ml distilled water and make the volume 1 litre in a volumetric flask).
2. 0.5N ferrous ammonium sulphate (Dissolve 196 g ferrous ammonium sulphate in 500 ml distilled water, add 20 ml of concentrated sulphuric acid and finally make the volume 1 litre). Freshly prepared, light green in colour.
3. Concentrated sulphuric acid.
4. Concentrated orthophosphoric acid (85% ) or Sodium fluoride.
5. Diphenylamine indicator: Dissolve 0.5 g of diphenylamine in a mixture of 100 ml concentrated sulphuric acid and 20 ml of distilled water.

*Procedure*

1. Accurately weigh 1 g of soil sample passed through 0.5mm sieve.
2. Place the weighed soil in a dry 500 ml conical flask.
3. Add exactly 10 ml of 1N potassium dichromate solution with the help of pipette and swirl the flask gently.
4. Rapidly add 20 ml of conc. Sulphuric acid with the help of measuring cylinder by directing steam into the suspension.

5. Swirl the flask and keep it on an asbestos sheet for 30 minutes.
6. Add about 200 ml of water.
7. After addition of 10 ml of orthophosphoric acid and 1 ml of diphenylamine indicator, titrate the contents against 0.5 N ferrous ammonium sulphate till colour flashes from blue violet to green. If the titre value is less than 4ml, repeat with less soil and if more than 17.0ml repeat with more soil.
8. Run a blank also (without soil).

*Calculations*

$$\text{Organic carbon (\%) in soil} = \frac{10\,(B-S)}{B} \times 0.003 \times \frac{100}{\text{Weight of soil (g)}}$$

Where, B and S are titre values for blank and soil sample, respectively.

***ii. Colorimetric method***

In the colorimetric method, the intensity of green colour of chromic acid obtained due to its reduction, is measured colorimetrically. The intensity is directly proportional to the amount of organic carbon present in the soil.

*Reagents*

1. Normal potassium dichromate solution (Dissolve 49.04 g of AR grade potassium dichromate in 500 ml distilled water and make the volume 1 litre in a volumetric flask).
2. Concentrated sulphuric acid.
3. Anhydrous sucrose (AR grade).

*Procedure*

1. Accurately weigh 1 g soil sample passed through 0.5mm sieve.
2. Place the weighed soil in a dry 100 ml conical flask.
3. Add 10 ml of 1N potassium dichromate solution with the help of pipette and add 20 ml of conc. Sulphuric acid with the help of measuring cylinder
4. Swirl the conical flask and keep it on an asbestos sheet for 30 minutes.
5. Decant about 15 to 20 ml of the supernatant liquid carefully into a centrifuge tube, retaining the soil particles in the conical flask as far as possible.

6. Centrifuge for about 5 minutes to ensure complete settling of soil particles, if any.
7. Carefully, decant the clear liquid into cuvette of colorimeter or spectrophotometer and read the colour intensity using red filter or at 660 nm wavelength.
8. Simultaneously, run a blank without soil.
9. Accurately weigh 5, 10, 15, 20 1nd 25 mg of anhydrous sucrose crystals into 100 ml conical and proceed as above on addition of dichromate, centrifugation and decantation.
10. Measure the intensity of blue colour by photoelectric colorimeter or spectrophotometer at 660 nm using a red filter. Prepare the standard curve by plotting quantity of sucrose against spectrophotometer reading.
11. Find out a factor to be multiplied by simple reading to get organic carbon (%) content of soil. The average value of this factor comes to be 0.0042 with little possible variation.

*Calculation*

Organic carbon (%) in soil = Spectrophotometer reading X 0.0042

**Caution:** The centrifuge may get damaged due to accidental spilling of chromic acid. Keeping the contents over night is suggested as an alternative.

## 4. Nitrogen

Nitrogen in soils occurs both in organic and inorganic forms,. The organic fraction, composed mostly of plant and microbial remains, is variable in composition. The inorganic phase of soil N is composed of ammonium ($NH_4^+$), nitrate ($NO_3^-$), and nitrite ($NO_2^-$) forms. Environmental (temperature and moisture) and management (fertilization, cropping, etc.) factors influence its dynamic relationship between organic and inorganic forms. The $NH_4$-N and $NO_3$-N forms are routinely measured in soil laboratories, as they reflect the extent of mineralization, and these are the forms of N taken up by plants.Nitrate-N content in soils is a good index for predicting N fertilizer need of crops. The organic-N fraction is a measure of the soil N reserve and its potential to release N for crop needs through mineralization. Thus, methods of N analysis vary depending on the N fractions.

Total soil N is generally measured after wet digestion using the well-known Kjeldahl procedure.In soil, nitrogen that is present in organic form is unavailable to plants. The available nitrogen in soil is that portion, which is present in mineral forms usually in the form of exchangeable ammonium and nitrate in the soil solution.

The procedure that are used to find the available nitrogen in a soil, include the determination of the total nitrogen content as correlated with plant response, measurement of the amount of nitrogen mineralized during a definite period of time, chemical extraction of inorganic forms of nitrogen present, determination of carbon dioxide produced when the soil is incubated with nitrogen free energy material and measurement of microbial growth.

Of the many methods, the one that appears to be used extensively in soil testing work for field crops,is the determination of the total content of organic matter and determination of the nitrogen mineralized during incubation of a sample of soil in the laboratory. The easily mineralisable nitrogen is widely used in soil testing laboratories and is considered to be a reliable index of nitrogen availability of aerobic soils, due to its high correlation with N uptake by plants.

***I) Total Nitrogen (Bremner and Mulvaney, 1982)***

The sample is digested with $H_2SO_4$ to convert organic N to $NH_4^+$-N. Highly refractory organic N compounds or compounds containing N-N or N-O linkages are not completely recovered by the Kjeldahl digestion; however, very little of the N in most soils is in this form. If soils contain high amounts of $NO_3^-$-N or $NO_2$ -N, then a pre-treatment must be carried out to include these forms of N. In undisturbed forest soils, negligible amounts of $NO_3^-$-N and $NO_2$ -N are present.

*Apparatus*

Digestion block: a 20-place block digester, distillation set for N analysis

*Reagents*

1. Concentrated $H_2SO_4$ (18 M), 96%, salicylic acid, sodium thiosulphate.
2. Catalyst Mixture ($K_2SO_4$ + $CuSO_4.5H_2O$ + Se powder), 100:10:1 w/w ratio

   Grind reagent-grade chemicals separately and mix. If caked, grind the mixture with a porcelain pestle and mortar to pass a 60mesh sieve (0.250 mm).
3. Sodium Hydroxide Solution (NaOH), 10 N:Dissolve 400 g sodium hydroxide in distilled water, transfer to a 1litre volumetric heavy walled Pyrex flask, let it cool, and bring to volume with distilled water.
4. Boric Acid Solution ($H_3BO_3$), saturated

*Procedure*

1. Accurately weigh 1.0 g of soil and transfer into a digestion tube
2. Add 10 ml concentrated $H_2SO_4$ and mix by swirling.
3. Add one spatula of digestion mixture, sodium thiosulphate and salicylic acid
4. Heat at 200°C in a digestion block until very black (about 30 minutes). To avoid acid irritation to the analyst, the digestion block must be loaded in a fume hood to ensure the removal of fumes and vapors released during digestion.
5. Increase heat to 300°C and heat for 30 minutes.
6. Raise the temperature to 375°C and heat for 30 minutes
7. Remove the digestion tubes from the block and allow to cool for 5 minutes.
8. Add about 50 ml water and mix well until the sample is in solution.
9. Transfer to a 500ml distillation flask and add 50 ml of distilled water.
10. Pipette 20 ml of 2% boric acid solution in 250 ml conical flask and add few drops of mixed indicator . Dip the end of the delivery tube into it.
11. Connect the flask to the distillation apparatus and start distillation.
12. With absorption of ammonia, pink colour of boric acid solution turns green, collect aproximately 50 ml of distillate.
13. Titrate the collected solution against 0.1 N sulphuric acid till the green colour disappears.
14. One reagent blank should be distlled and titrated as above.

*Calculations*

$$\text{Total N (kg N/ha)} = \frac{14x\ (S-B) \times 0.1 \times 2.24 \times 1000}{\text{Weight of soil taken (g)}}$$

Where, S and B are titre values for sample and blank, respectively.

***II) Available N (easily oxidizable/mineralizable N, Subbiah and Asija, 1956)***

*Principle*

A known weight of soil is mixed with excess of alkaline $KMnO_4$. In alkaline medium, $KMnO_4$is a mild oxidizing agent. Organic matter, present in the soil, is oxidized by the nascent oxygen liberated by $KMnO_4$ in the presence of NaOH. Ammonia thus formed, is liberated by steam distillation and in a known volume of standard sulphuric acid (N/50 $H_2SO_4$). The excess acid is titrated with standard alkali (N/50 NaOH) using methyl red as the indicator.

The evolved ammonia may also be absorbed in boric acid indicator solution and the absorbed ammonia can be determined by titration with standard acid (*N/50* $H_2SO_4$). The reactions are stated below.

**(A) Oxidation :**

$$2\ KMnO_4 + H_2O \xrightarrow{\text{Alkaline}} 2MnO2 + 2KOH + 30^-$$

$$RCH.NH2.COOH + O^- \xrightarrow{\text{Medium}} R.CO.COOH + NH_3$$

**(B) Distillation :**

$$NH_3 + H_2O \longrightarrow NH_4OH$$

**(C) Absorption in $H_2SO_4$ and titration with NaOH**

$$2NH_4OH + H_2SO_4 \longrightarrow (NH_4)_2SO_4 + 2H_2O$$

**(D) Absorption in $H_3BO_3$ and titration with $H_2SO_4$**

$$NH_4OH + H_3BO_3 \longrightarrow NH_4\ H_2BO_3 + H_2O$$

$$NH_4\ H_2BO_3 + H_2SO_4 \longrightarrow (NH_4)SO_4 + 2\ H_3BO_3$$

*(A) Apparatus*

Steam distillation set, measuring cylinder, pipette, burette, conical flask, heater or burner, red litmus paper.

*(B) Reagents:*

1. Potassium permanganate solution (0.32%): Dissolve 3.2g potassium permanganate in 1 litre volumetric flask with distilled water and make up the volume to one litre.
2. Sodium hydroxide solution (2.5%): Dissolve 25 g sodium hydroxide in 1 litre volumetric flask with distilled water and dilute to one litre.

3. Mixed indicator: Take 0.099 g of bromocresol green and 0.066 g of methyl red in 100ml of ethanol.
4. Boric acid (2% boric acid): Dissolve 20 g of boric acid in 500 ml warm distilled water and make up the volume to 1litre. Add 20 ml of this mixed indicator to each litre of 2% boric acid solution.
5. 0.1N Potassium hydrogen phthalate: Dissolve 20.422 g of the salt in water and dilute to 1 litre.
6. 0.1 N Sodium hydroxide solution: Dissolve 4 g NaOH in water and dilute to 1 litre in a1litre volumetric flask. Standardize this against 0.1 N Potassium hydrogen phthalate solution.
7. Sulphuric acid solution: Add 2.8 ml of conc. $H_2SO_4$ to 990 ml distilled water and make up the volume to 1 litre, to obtain approximately 0.1N $H_2SO_4$. Standardize this against 0.1 N NaOH solution. Then, prepare 0.02 N $H_2SO_4$ by diluting a suitable volume approximately five times depending upon the volume of NaOH consumed with distilled water.

*Procedure*

1. Place 20 g of soil in 800 ml kjeldahl flask.
2. Add 10 ml distilled water and wash down the soil in the flask.
3. Add 100 ml of 0.32% potassium permanganate solution.
4. Add 2-3 ml paraffin liquid to avoid contact with upper part of the neck of the flask and a few glass beads or broken glass rod.
5. Pipette 20 ml of 2% boric acid solution in 250 ml conical flask and dip the receiver end of the delivery tube into it.
6. Pour 100ml of 2.5% sodium hydroxide into the distillation flask and cork it immediately. Run tap water in the condenser.
7. Connect the flask to the distillation apparatus and start distillation by switching on the heater. Continue distillation until about 100 ml of distillate is collected.
8. First remove the conical flask containing distillate and switch off the heater to avoid back suction.
9. Titrate the distillate against 0.02 N $H_2SO_4$ taken in burette until pink colour starts appearing.
10. Run a blank without soil

11. Remove the Kjeldahl flask after cooling and drain the contents. Collect the glass beads for reuse.

*Calculations*

$$\textbf{Available N (kg N/ha) in soil} = \frac{\textbf{(S-B)} \times \textbf{0.00028} \times \textbf{10}^6 \times \textbf{2.24}}{\textbf{20}}$$

Where, B and S are titre values for blank and soil sample, respectively.

The factor 0.00028 is arrived by considering the following equations[2]

$NH_4OH + H_2SO_4 = (NH_4)_2\ SO_4 + 2H_2O$

1 ml 1 N $H_2SO_4 \equiv$ 14 mg N

Hence, 1 ml of 0.02 N $H_2SO_4$ = 0.00028 g N

### *III. Mineral -N*

Mineral-N is determined by using 2 M KCl, as the extracting solution in a 1:5-10 (soil: water) ratio. Ammonium ($NH_4^+$) and nitrate ($NO_3^-$) plus nitrite ($NO_2^-$) are determined by steam distillation of ammonia ($NH_3$), using MgO for $NH_4^+$ and further 2nd stage distillation by addition of Devarda's Alloy for $NO_3$- (Bremner and Keeney, 1965). The distillate is collected in saturated $H_3BO_3$ and titrated to pH 5.0 with dilute $H_2SO_4$. This method determines dissolved and adsorbed forms of $NH_4^+$, $NO_3^-$ and $NO_2^-$ in soils. The sum of $NH_4^+$ and $NO_3^-$ determined by this method is referred to as Mineral-N (Keeney and Nelson, 1982; Buresh, et al., 1982).

Ammonium is held in an exchangeable form in soils in the same manner as exchangeable metallic cations. Fixed or nonexchangeable $NH_4^+$make up a significant portion of soil N; however, fixed $NH_4$ is defined as the $NH_4$ in soil, that cannot be replaced by a neutral K salt solution (Keeney and Nelson, 1982). Exchangeable $NH_4^+$ is extracted by shaking with 2.0 M KCl. Nitrate is water-soluble, and hence can also be extracted by the same 2.0 M KCl extract. Nitrite is seldom present in detectable amounts in soil and therefore is usually not determined.

*Reagents*

A. Potassium Chloride Solution (KCl), 2 M; Dissolve 150 g reagent potassium chloride in distilled water, and bring to 1 litre.

B. Magnesium Oxide (MgO), powder: Heat heavy magnesium oxide in a muffle furnace at 600 - 700°C for 2 hours, and cool in a desiccator containing KOH pellets, and store in a tightly stoppered bottle.

C. Devarda's Alloy (50 Cu: 45 Al: 5 Zn) Ball-mill reagent-grade Devarda's Alloy until the product will pass a 100-mesh sieve (0.1 50-mm) and at least 75% will pass a 300-mesh sieve (0.050-mm).

D. Boric Acid ($H_3BO_3$), saturated solution.

E. Sulfuric Acid Solution ($H_2SO_4$), 0.01 N

F. Standard Stock Solution

- Dry reagent-grade ammonium sulphate [$(NH_4)_2SO_4$], and potassium nitrate ($KNO_3$) in an oven at 100°C for 2 hours, cool in a desiccator, and store in a tightly stoppered bottle.
- Dissolve 5.6605 g ammonium sulphate and 8.6624 g potassium nitrate in distilled water, and transfer to a 1 litre volumetric flask, mix well, and make up the volume. This solution contains 1.2 g $NH_4$-N, and 1.2 g $NO_3$-N per litre (Stock Solution).
- Prepare a standard solution from the stock solution: dilute 50 mL stock solution to 1 litre volume by adding 2 M potassium chloride solution (diluted stock solution).
- A 20-ml aliquot of diluted Stock Solution contains 1.2 mg $NH_4$-N and 1.2 mg $NO_3$-N.

*Procedure*

1. Weigh 30 g air-dry soil (2 mm) into a 250-ml Erlenmeyer flask and add 150 ml 2 M potassium chloride solution (1:5 soil: solution ratio).
2. Stopper flasks, shake for 1 hour on an orbital shaker at 200 - 300 rpm, and filter suspensions using Whatman No. 42 filter paper.
3. Take 50 ml of aliquot in a distillation flask.
4. Pipette 20 ml of 2% boric acid solution in 250 ml conical flask and add few drops of mixed indicator to it. Dip the receiver end of the delivery tube into it.
5. Add 5 g of MgO into the distillation flask and cork it immediately.
6. Connect the flask to the distillation apparatus and start distillation.
7. With absorption of ammonia, pink colour of boric acid solution turns green ,collect approximately 100 ml of distillate.
8. Titrate the collected solution against 0.01 N sulphuric acid till the green colour disappears.

9. One reagent blank is to be made for the final calculations.
10. To determine $NO_3$-N (plus $NO_2$-N) in the same extract, add 0.2 g Devarda's alloy with a calibrated spoon to the same distillation flask.
11. Attach flask to distillation unit with a clamp, and start distilling. Further proceed as for ammonium-N.
12. Between different samples, steam out the distillations. Disconnect distillation flasks containing the KCl extracts, and attach a 100ml empty distillation flask to the distillation unit, and place a 100ml empty beaker underneath the condenser tip, turn off cooling water supply (drain the water from the condenser jacket), and steam out for 90 seconds. Steaming out is done only between different samples, not between distillation for ammonium (MgO) and nitrate (Devarda's alloy) in the same sample.
13. Each distillation should contain at least two standards and two blanks, i.e., 2 M KCl extracts with no soil added (reagent blank).

If possible, mineral-N should be determined in field-moist soil, immediately after sampling. However, analytical results should be expressed on an oven-dry soil basis. If the analysis cannot be done immediately after sampling, soil samples may be kept in a cool place.

## 5. Phosphorus

In soil, phosphorus exists in the form of various types of orthophosphates. A very small fraction of these is available to plants at a given time. Available P content of soil consists mainly of Ca-, Al- and Fe-P. In neutral, alkaline and calcareous soils, Ca-P is dominant; in acid soils Al- and Fe-P are dominant. Two types of extraction methods viz., Olsen's method and Bray &Krutz's method are popularly adopted for alkaline and acid soils, respectively, in soil testing laboratories.

### *I. Determination of Soil Available Phosphorus (Olsen et al., 1954)*

This method is suitable for soils having pH greater than 6.5. It has been found that soils, which are slightly acid, neutral, alkaline and calcarcous are highly correlated with the crop response to phosphate fertilization based on the P estimation by Olsen's Method.

In this method, the available phosphorus of soil is extracted by shaking the soil with 0.5M sodium bicarbonate solution adjusted to pH 8.5 in 1 : 20 ratio for half an hour in presence of Darco-G60 carbon, which absorbs the dispersed organic colloids and thus, helps to give a clear extract. Phosphorus, thus extracted in the suspension is filtered. Phosphorus in the filtered extract is treated with ammonium molydate, which forms yellow coloured phosphomolybdate complex, which is reduced by stannous chloride resulting in the formation of blue coloured reduced phosphomolybdate complex. The intensity of blue colour, which is proportional to the amount of phosphorus, is determined colorimetrically by chlorostannous reduced molybdophosphoric blue colour method in hydrochloric acid system.

***The reactions are stated below:***

*A) Extraction*

(a) Soil Colloid Phosphate + $NaHCO_3 \rightarrow$ Soil Colloid $HCO_3$ + Sodium phosphate

(b) $Ca_3(PO_4)_2 + 6\ NaHCO_3 \rightarrow 3Ca\ (HCO_3)_2 + 2\ Na_3PO_4$

$3Ca\ (HCO_3)_2 \rightarrow 3CaCO_3 + 3\ H_2CO_3$

$2\ Na_3PO_4 + 3\ H_2CO_3 \rightarrow 2H_3PO_4 + 3\ Na_2CO_3$

$Ca_3(PO_4)_2 + 6\ NaHCO_3 \rightarrow 3CaCO_3 + 2\ H_3PO_4 + 3\ Na_2CO_3$

*B) Development of Colour*

$(NH_4)_6\ Mo_7O_{24}.4H_2O + 6\ HCl \rightarrow 7H_2MoO_4 + 6NH_4Cl$
(Ammonium molybdate) (Molybdenic acid)

$H_3PO_4 + 12\ H_2\ MoO_4 \rightarrow H_3P(Mo_3O_{10})_4 + 12\ H_2O$
(Phosphate) (Yellow coloured Phosphomolybdate complex)

$H_3P\ (Mo_3O_{10})_4 + SnCl_2$ reduction $\rightarrow$ Blue coloured reduced (Phosphomolybdate complex)

*A) Equipment and other materials*

Photoelectric colorimeter or spectrophotometer, shaker, pH meter, conical flask (100 ml), pipette (5 ml), volumetric flask (100 ml, 25ml), funnel, Whatman No. 1 filter paper.

*(B) Reagents*

**(i) Sodium bicarbonate solution (0.5*M*):** 42.0g of $NaHCO_3$ is dissolved in about 500ml distilled water. The pH of the solution is adjusted to 8.5 with 10% NaOH solution or dilute HCl and the volume is made upto 1 litre.

**(ii) Darco-G60 carbon:** The Darco-G60 carbon is repeatedly washed with 0.5 *N* $NaHCO_3$ solutions till it is free from phosphorus. It is dried and stored.

**(iii) 1.5% Ammonium molybdate solution:** 15g of ammonium molybdate is dissolved in 400ml of warm water at $60^0C$, the solution is cooled, (filtered, if turbidity exists) and mixed with 400ml of 10N HCl and the volume is made upto 1 litre.

**(iv) Stannous chloride solution:**

**(a)** ***Stock solution*:**Weigh 10 g of pure $SnCl_2.2H_2O$ crystals in a 100 ml glass beaker. Add 25ml of conc. HCl and dissolve by heating. Cool, transfer to an amber coloured bottle and store pin a dark place after adding a small piece of Zn metal (AR grade) to prevent oxidation.

**(b)** ***Dilute solution*:** 0.5 ml of the stannous chloride stock solution is added to 66ml of water. This must be prepared fresh for each set of determination.

**(v) Standard P solution:** 0.439 g of AR grade potassium dihydrogen phosphate ($KH_2PO_4$) which is dried in an oven at $60^0C$ for an hour is taken in one litre volumetric flask and dissolved in about 500ml distilled water. About 25ml of approx. $7NH_2SO_4$ is added to it and the volume is made up to 1 litre with distilled water. This gives 100 ppm (100 mg/litre) stock solution of P.

**(vi) 2.5ppm P solution:** 2.5ml of 100ppm P solution is taken in a 100ml volumetric flask and diluted to 100ml with distilled water.

*Preparation of standard curve for P*

Take six numbers volumetric flasks of 25 ml of in which 0, 1,2,3,4 and 5ml of 2.5 ppm P solution is added. Add 5 ml of the extractant (0.5 *M* $NaHCO_3$) and 5ml of ammonium molybdate solution and shake slowly. Add 1 ml of freshly diluted $SnCl_2$ solution and make up to 25ml with distilled water, shake slowly and allow to develop blue colour. Measure the intensity of blue colour by photoelectric colorimeter or spectrophotometer at 660nm using a red filter. Prepare the standard curve by plotting concentrations of P in ppm against % transmittance or optical density or colorimeter reading on a semilog graph paper. If straight line is obtained, a factor for each reading is found out.

*Procedure*

1. Weigh 2.5 g of soil in a 100 ml conical flask.
2. Add a pinch of Darco-G 60 carbonand 50ml of sodium bicarbonate (0.5M) extraction solution (pH 8.5)

3. Shake for 30 minutes on a mechanical shaker.
4. Similarly run a blank without soil.
5. Filter through Whatman No.1 filter paper.
6. Place 5ml of aliquot in a 25 ml volumetric flask.
7. Gradually add (drop by drop) 5 ml of 1.5% ammonium molybdate solution and shake slowly to drive out the $CO_2$ evolved.
8. When frothing completely ceases, add distilled water, bring the volume to 22 ml (approx.)
9. Add 1 ml of freshly diluted $SnCl_2$ solution, shake a little and make up the volume to 25 ml.
10. Read the blue colour intensity at 660 nm wavelength (red filter).
11. Run a blank without soil under identical manner.
12. The colorimeter reading or absorbance of the sample are located on standard curve and corresponding concentrations of P is calculated.

*Calculations*

$$\text{Available P (kg P/ha) in soil} = \frac{Q \times V \times 2.24 \times 10^6}{A \times S \times 10^6} = \frac{Q \times V \times 2.24}{A \times S}$$

Where, Q = quantity of P in µg obtained from graph V = volume of extracting reagent used (ml) A = volume of aliquot used for colour development (ml) and S = weight of soil sample (g)

*Precautions*

(1) If the filtrate is not colourless, more Darco-G 60 carbon should be added, shaken and filtered.

(2) Dilute $SnCl_2$ should be prepared afresh before use.

(3) Development of blue colour is completed after about 10 minutes, but it starts fainting after 15 minutes. So, reading should be taken within 10 to 15 minutes of addition of $SnCl_2$.

(4) If intensity of colour is too high to be read, smaller amount of the aliquot is to be taken to redevelop the colour.

(5) For accurate results,clean the glassware with sulphuric acid – dichromate solution and wash them thoroughly with tap water followed by rinsing with distilled water.

### *II. Determination Available Phosphorus in Soil by Bray's P-1 (Bray and Kurtz, 1945)*

This method is mainly suitable for soils, which have pH 6.5 or less. It has been found that soils, whose acidity is moderate or more are highly correlated with the crop response to phosphate fertilizer based on P estimation by Bray's P1 method.

In this method, the available phosphorus of soil is extracted by shaking the soil with ammonium fluoride- hydrochloric acid solution in 1: 10 ratio. The $NH_4F$ – HCl solution dissolves the fraction of P, that are considered to be available to plants. As ammonium fluoride complexes iron and aluminum ions present in the soil solution of an acid soil, so, the P held by $Fe^{3+}$ and $Al^{3+}$ ions are released. Thus, the ammonium fluoride-hydrochloric acid solution helps to extract easily acid soluble forms of P (mainly calcium phosphate) and a part of iron and aluminium phosphate.

The probable reactions in acid soil are stated below:

$$3NH_4F + 3HF + AlPO_4 \longrightarrow H_3PO_4 + (NH_4)_3\,AlF_6$$

$$3NH_4F + 3HF + FePO_4 \longrightarrow H_3PO_4 + (NH_4)_3\,FeF_6$$

These $AlPO_4$ and $FePO_4$ represent various hydrated and hydroxyl phosphates of Al and Fe, including the adsorbed or precipitated surface layers of oxides of iron and aluminum and alumino silicates.

HCl helps in the dissolution of more active phosphate and prevents the precipitation of P (as calcium phosphate), which has been dissolved by $NH_4F$.

Phosphorus in the filtered extract is treated with ammonium molybdate, which forms yellow coloured phospho-molybdate complex, which is reduced by stannous chloride resulting formation of blue coloured reduced phospho-molybdate complex.

The Reactions are stated below:

$$(NH_4)_6\,Mo_7O_{24}.4H_2O + 6\,HCl \longrightarrow 7H_2OMoO_4 + 6NH_4Cl$$

$$H_3PO_4 + 12\,H_2MoO_4 \longrightarrow H_3P\,(Mo_3O_{10})_4 + 12\,H_2O$$

#### *A) Reagents*

i. **Bray's P-1 extractant:** Dissolve 1.110 g AR grade $NH_4F$ in one litre of 0.025 N HCl.

ii. **1.5% Ammonium molybdate solution:** 15 g of ammonium molybdate is dissolved in 400 ml of warm water at $60^0C$, the solution is cooled, (filtered, if turbidity exists) and mixed with 400 ml of 10N HCl and the volume is made upto 1 litre.

**iii. Stannous chloride solution**

a. **Stock solution:** Weigh 10 g of pure $SnCl_2.2H_2O$ crystals in a 100 ml glass beaker. Add 25 ml of conc. HCl and dissolve by heating. Cool, transfer to an amber coloured bottle and store pin a dark place after adding a small piece of Zn metal (AR grade) to prevent oxidation.

b. **Dilute solution:** 0.5 ml stannous chloride stock solution is added to 66ml of water. This must be prepared fresh for each set of determination.

**iv. Standard P solution:** 0.439 g of AR grade potassium dihydrogen phosphate ($KH_2PO_4$), which is dried in an oven at $60^0C$ for an hour is taken in one litre volumetric flask and dissolved in about 500 ml of distilled water. About 25ml of approx. 7N $H_2SO_4$ is added to it and the volume is made up to 1 litre with distilled water. This gives 100 ppm (100 mg/litre) stock solution of P.

**v. 2.5 ppm P solution:** 2.5 ml of 100 ppm P solution is taken in a 100 ml volumetric flask and diluted to 100 ml with distilled water.

***Preparation of standard curve for P***

Take six numbers volumetric flasks of 25 ml of in which 0, 1, 2, 3, 4 and 5ml of 2.5 ppm P solution is to be added. Add 5 ml of the extractant and 4ml of 1.5% ammonium molybdate. Make up to 25ml with distilled water, shake slowly and allow to develop blue colour. Measure the intensity of blue colour by photoelectric colorimeter or spectrophotometer at 660nm using a red filter. Prepare the standard curve by plotting concentrations of P in ppm against % transmittance or optical density or colorimeter reading on a semilog graph paper. If straight line is obtained, a factor for each reading is found out.

*Procedure*

1. Weigh 5.0 g of soil in a 100 ml conical flask.
2. Add 50 ml of extracting solution (soil-to-solution ratio of 1:10).
3. Shake for 5 minutes on a mechanical shaker.
4. Filter through Whatman No.1 filter paper.
5. Place 5ml of aliquot in a 25 ml volumetric flask.
6. Add 5 ml of 1.5% ammonium molybdate solution
7. Add 1 ml of freshly diluted $SnCl_2$ solution, shake a little and make up the volume to 25 ml.

8. Read the blue colour intensity at 660 nm wavelength (red filter).
9. Run a blank without soil under identical manner.

*Calculations*

$$\text{Available P (kg P/ha) in soil} = \frac{Q \times V \times 2.24 \times 10^6}{A \times S \times 10^6} = \frac{Q \times V \times 2.24}{A \times S}$$

Where, Q = quantity of P in µg obtained from graph

V = volume of extracting reagent used (ml)

A = volume of aliquot used for colour development (ml) and

S = weight of soil sample (g)

***Ascorbic acid method of P estimation***

A method based on the use of ascorbic acid instead of stannous chloride gives more stable blue colour and is, therefore, preferred over stannous chloride

**a.** **Molybdate-tartarate solution:** Dissolve 12 g of ammonium molybdate in about 250 ml of distilled water to get solution "A". Prepare solution "B" by dissolving 0.291 g of antimony potassium tartarate in 100 ml of distilled water. Prepare one litre of 5 N $H_2SO_4$ and add solutions A and B to it. Mix thoroughly and make the volume to 2 litres with distilled water.

b. **Ascorbic acid solution:** Dissolve 1.056 g of ascorbic acid in 200 ml of the molybdate-tartarate solution and mix well. Prepare it fresh as and when required.

## 6. Potassium

### *I. Determination of Available Potassium (Hanway&Heidel, 1952) in soil by Flame Photometer Method*

Available potassium (K) refers to exchangeable potassium adsorbed on soil colloidal surfaces plus water soluble potassium present in the soil solution. This is only 1 to 2% of the total soil potassium. Although about 90% of available potassium is in exchangeable form, soluble potassium is readily absorbed by higher plants. The neutral normal ammonium acetate extract contains both water soluble and exchangeable potassium. As potassium extracted by this method is found to be a suitable index of potassium availability in most of the soils on the basis of crop response study, so potassium extracted by this method is considered as available potassium.

In this method, water soluble and the exchangeable potassium of the soil is extracted by shaking the soil with neutral normal ammonium acetate in 1 : 5 soil: solution ratio. During equilibrium, ammonium ions exchange with the exchangeable potassium ions along with other exchangeable cations. After equilibrium, the suspension is filtered through Whatman no. 1 filter paper. The potassium content of the filtrate is determined with a flame photometer.

*(A) Equipments and other materials:*

Balance, Flame photometer, pH meter, 150 ml conical flask, funnel, 250ml beaker, 5,10 and 20ml pipette, 1000 and 100ml volumetric flask, Whatman No. 1 filter paper.

*(B) Reagents*

**(1) Normal neutral ammonium acetate solution**

Dissolve77.08 g of ammonium acetate in about 500 ml distilled water and make up the volume to less than 1 litre. Adjust the pH of the solution to 7.0 by addition (drop by drop) of glacial acetic acid or ammonia solution.

Or

Take 800 ml of distilled water and add 57 ml of glacial acetic acid and 68 ml of ammonia solution (Sp. gr. 0.91) followed by dilution to 1 litre. Adjust the pH to 7.0 after cooling.

**(2) Standard potassium solution**

(a) Stock solution: Dissolve 1.908g of pure AR grade KCl (dried at $60^0$C for 1 hour) in distilled water and make up the to one litre in a volumetric flask. This solution contains 1000 ppm K. It is treated as stock solution of K.

(b) 100 ppm standard Ksolution: Take 10 ml of stock solution in a 100 ml volumetric flask and make up the volume upto the mark with distilled water.

(c) Standard curve : Take 0,5,10,15,20,30 and 40 ml of 100ppm stock solution in 100ml volumetric flasks and make the volume up to the mark with extracting solution (i.e. neutral normal ammonium acetate). The solutions contain 0, 5, 10, 15, 20, 30 and 40 ppm K respectively. Potassium standards should be prepared fresh after every 2-3 weeks.

*Preparation of standard curve*

Insert the K filter and regulate the appropriate gas, pressure and light the burner and adjust the air pressure. Set the flame photometer reading at 0 for blank ($NH_4OAC$) and at 100 for 40 ppm of K solutions alternatively, till both values are obtained without any adjustment. Take the reading for standards and plot a graph by taking flame photometer reading in Y axis and concentration of K in X-axis.

*Procedure*

1. Weigh 5.0 g of 2 mm air dry soil into a 100 ml conical flask.
2. Add 25 ml of extracting solution (soil-to-solution ratio of 1:5).
3. Shake for 5 minutes in a mechanical shaker.
4. Filter the suspension using Whatman No. 1 filter paper.
5. Atomize the filtrate in the flame photometer in which 0 and 100 have been set with blank and 40 ppm K solutions, respectively.
6. Record flame photometer reading
7. Calculate concentration of K from the plotted graph or standard curve.

*Precautions*

(i) If K is determined without destroying $NH_4OAC$, standard solutions must contain $NH_4OAC$ in like concentration.

(ii) When K is determined in sodic soils, it is necessary to use internal standard to overcome the interference of the Na.

(iii) The standards and samples must be kept free from any suspended material to prevent the clogging of the nebulizer and capillary.

(iv) Highly concentrated solution should not be used for determination by flame photometer.

(v) The flame should not be left unattended.

(vi) The pressure tubing of the fuel gas should be replaced periodically.

*Calculations*

$$\text{Available K (kg K/ha) in soil} = C \times \frac{25}{5} \times 2.24 = C \times 11.2$$

Where, C = Concentration (mg/litre) of K in the sample filtrate obtained on X- axis against the reading

***II. Boiling Nitric Acid Extraction for Potassium***

In this method, water soluble, exchangeable and non-exchangeable potassium of the soil is extracted by boiling with 1N $HNO_3$ in 1: 10 soil: extractant.

Subtract the quantity of K obtained by neutral normal $NH_4OAC$ method from that in 1N $HNO_3$ extract, to obtain data for release of K from non-exchangeable forms.

*Procedure*

1. Weigh 2.5 g of soil (0.1-g accuracy) in a 100 ml conical flask.
2. Add 25 ml of 1.0 N $HNO_3$ (soil-to-solution ratio of 1:10).
3. Place the flask on a ring stand over gas burner.
4. When boiling starts, reduce the flame and boil for 10 minutes.
5. Remove the flask, pour the contents into a filter and collect the filtrate in a 100 ml volumetric flask.
6. Wash the soil with four 15 ml portions of 0.1N $HNO_3$.
7. Cool the solution, make up the volume to mark.
8. Determine the K content with a flame photometer using appropriate standards made in the same $HNO_3$ as in sample.
9. Atomize the filtrate in the flame photometer in which 0 and 100 have been set with blank and 40 ppm K solutions respectively.
10. Record flame photometer reading and calculate concentration of K from the plotted graph or standard curve.

1N $HNO_3$ boiling K (kg K/ha) = ppm from graph x 2.24 x 40

Non-Exchangeable K (kg K/ha) = 1N $HNO_3$ boiling K - Available K (kg K/ha)

**7. Determination of Available Sulphur (Williams and Steinberg, 1969)**

Adsorbed and soil solution $SO^{2-}{}_4$ ions constitute the available sulphur in soils. Both 0.15% $CaCl_2$and 500ppm P solutions of Ca & K are used to replace adsorbed $SO^{2-}{}_4$ ions. Use of Ca salts have the advantage over those of Na or K, as Ca prevents deflocculation and leads to easy filtration. The $SO^{2-}{}_4$ extract is estimated turdidimetrically using a Spectrophotometer.

***(A) Equipment and Apparatus***

Photoelectric colorimeter or spectro photometer, mechanical shaker, Whatman No. 42 filter paper, Erlenmeyer flask, volumetric flask, pipette.

***B) Reagents***

**(a) $CaCl_2$ dihydrate ($CaCl_2.2H_2O$) :** 0.15%: Dissolve 1.5 g $CaCl_2.2H_2O$

is in about 500 ml water and make up the volume to 1 litre with distilled water in volumetric flask.

**(b) $BaCl_2$ dihydrate ($BaCl_2.2H_2O$):** Grind $BaCl_2.2H_2O$ crystals and pass the particles through 30 mesh sieve and retain on 60 mesh sieve and store in a clean bottle.

**(c) Gum acacia solution (0.25):** Dissolve 0.25 g of gum acacia in 50 ml of warm water and make up the volume to 100 ml in a volumetric flask.

**(d) Standard sulphate solution:** Dissolve 0.5434 g of AR grade $K_2SO_4$ in 100 ml distilled and dilute to 1 litre. This contains 100 µg S per ml of solution (100ppm S), prepare working standard solution of 10 ppm S solution by diluting 10 times with distilled water.

*Preparation of standard curve*

In a series of 25 ml volumetric flasks, pipette out 0, 2.5, 5.0, 7.5,10 and 12.5 ml of 10 ppm S solution, which will correspond to 0, 1, 2,3,4 and 5 ppm S standard solution, respectively. Add 10 ml of $CaCl_2$ solution, followed by approx. 1 g of barium chloride to each of the flask and swirl slowly to dissolve the crystals. Add 1 ml of 0.25 % gum acacia solution, make up the volume with distilled water and shake well manually. Within 5-10 minutes of development of turbidity (white colour), absorbance is read at 440 nm on a spectrophotometer or a colorimeter using blue filter.

*Procedure*

1. Weigh 10 g of soil in 150 ml conical flask.
2. Add 50 ml of 0.15% $CaCl_2$solution and shake for half an hour on a rotary shaker.
3. Filter the solution in a reagent bottle, through Whatman No. 42 filter paper.
4. Pipette out 10 ml of the aliquot in 25 ml volumetric flask, to it add approx. 1 g of barium chloride crystals and swirl to dissolve the crystals.
5. Add 1ml of 0.25% gum acacia solution, make up the volume up to 25 ml, shake well manually, turbidity (white colour) develops within 5-10 minutes.
6. Read at 440 nm on a spectrophotometer or a colorimeter using blue filter.
7. Run a blank (without soil) by following the above steps.

8. Derive the concentration of S in soil extracts from the standard curve

$$\text{Available S (mg/kg) in soil} = R \times \frac{50}{10} \times \frac{110}{10}$$

NB: In soils with low available S, 20 ml of aliquot in stead of 10 ml can be taken for development of turbidity.

## 8. Determination of Available Cationic Micronutrients (Fe, Cu, Mn, Zn, Co & Ni)

Micronutrient soil tests measure the magnitude of micronutrient pools in soil, that are related to plant availability. The chemicals that are used for extracting micronutrients from soil samples belong to different distinct groups, such as (i) water, (ii) dilute acids, (iii) chelating or complexing agents, and (iv) neutral buffered and unbuffered salts. These chemicals are used either singly or in combination for increasing their effectiveness in extracting micronutrients from soil. Wide range of chemical reagents have been used for determination of available Zn, Cu, Fe and Mn, while hot water, $NH_4OAc$ (pH 7.0) and Mehlich 3 extractant have been used for extracting available B and ammonium oxalate ($(NH_4)_2C_2O_4$) have been commonly employed for extraction of plant available molybdenum (Mo) in soils. There has been a general trend towards the use of one or a few soil test procedures, which can be used for several micronutrients and also potentially toxic elements in one extraction. On a global scale, DTPA is now the most widely used soil extractant for cationic micronutrients, but EDTA, hydrochloric acid, ammonium bicarbonate–DTPA and Mehlich 1 and Mehlich 3 tests are also relatively popular.

***Principle (DTPA extractable cationic micronutrients) (Lindsay and Norwell, 1978)***

In this method, the soil is shaken with a buffered extracting solution containing DTPA (diethylene triamine penta-acetic acid, (Lindsay and Norwell, 1978), $CaCl_2$ and TEA (triethanolamine). DTPA is buffered at pH 7.3 by TEA. Excessive dissolution of $CaCO_3$ especially in calcareous soil is prevented by inclusion of soluble calcium salt like $CaCl_2$,which also stabilize the pH of the extractant. DTPA is a mild chelating agent and has the ability to chelate Zn, Cu, Fe, Co, Ni and Mn in soil solution and forms soluble complexes. When the extractant solution is added to the soil, the ionic concentration of $Ca^{++}$ and $Mg^{++}$ in soil solution is increased due to exchange of protonated TEA with the $Ca^{++}$ and $Mg^{++}$ ions in the exchange sites, resulting suppression of the dissolution of $CaCO_3$.

The amount of soluble elements in the extracts is determined by the Atomic Absorption Spectrophotometer, wherein the extracted element is converted into an atomic vapour or free atoms by a flame. A light source provided by a "hollow cathode lamp" is used to excite the free atoms formed in the flame by the absorption of the electromagnetic radiation. As the absorption of light by the vaporized samples is related to the concentration of the element, the decrease in energy is measured following the Lambert-Beer law, i.e. the absorbance is proportional to the number of free atoms in the ground.

***Materials required***

*A) Apparatus and other material*

100ml conical flasks, 50, 100, 1000ml volumetric flasks, pipette, 2, 50 and 500ml measuring cylinders, Whatman No. 42 filter paper, horizontal shaker, Atomic Absorption Spectrophotometer, pH meter, double glass distilled water.

*Reagents*

1. **Dilute HCl:** Analytic reagent grade. Concentrated HCl is diluted to 5 times with double glass distilled water.
2. **DTPA Extractant:** Dissolve 1.967 g of analytical reagent (AR) grade DTPA (Diethylene triamine penta acetic acid) in 250 ml of double glass distilled water, allow sufficient time to dissolve DTPA. Add 1.470g of AR grade $CaCl_2$. 2 $H_2O$ and dilute to about 200ml by adding double glass distilled water, add 13.3 ml of TEA (Triethanolamine). Adjust the pH of the solution to 7.3 ± 0.05 by additing dilute HCl with stirring. Make up the mark up to one litre by adding double glass distilled water in a 1litre volumetric flask. This reagent has 0.005 M DTPA, 0.01 *M* $CaCl_2.2H_2O$ and 0.01 M TEA and the solution is stable for several months.
3. **Standard Stock solutions for Zn, Cu, Fe and Mn:** 100 ppm stock solution of each element is prepared in the following way :

| Metal | Atomic weight | Compound/ element used | Molecular weight (g) | Other chemicals required | Wt. of compound/ element per litre of solution (g) |
|---|---|---|---|---|---|
| Zn | 65.38 | $ZnSO_4.7H_2O$ | 287.56 | - | 4.398 |
| | | Pure Zn metal | 65.38 | 10 ml diluted HCl (1:1 with DDW) | 1.0 |
| Cu | 63.54 | $CuSO_4.5H_2O$ | 249.69 | - | 3.929 |
| | | Pure Cu metal | 63.38 | 50 ml diluted HCl (1:1 with DDW) | 1.0 |
| Fe | 55.85 | $FeSO_4.7H_2O$ | 278.02 | - | 4.977 |
| | | Pure Fe metal | 55.85 | 50 ml diluted $HNO_3$ (1:1 with DDW) | 1.0 |
| Mn | 54.94 | $MnCl_2.4H_2O$ | 197.91 | - | 3.602 |
| | | $MnO_2$ | 86.9 | 50 ml diluted $HNO_3$ (1:1 with DDW) | 1.583 |

4. **Standard 50 ppm solution:** Take50ml of 1000 ppm solution in a 100ml volumetric flask and dilute the solution up to the mark with double glass distilled water.

5. **Standard working solutions:** Take 0,0.5, 1.5, 2.0, 2.5 and 5.0 ml of 50 ppm solutions in 6 numbered 50ml volumetric flask. Make up the mark with the extracting solution. The working solutions contains 0, 0.5, 1.0, 1.5, 2.0, 2.5 and 5.0 ppm. The standard solutions vary from element to element, which may be verified from the instruction manual of the instrument.

*Procedure*

1. Weigh 10 g of soil sample in 100 ml conical flask
2. Add 20 ml of the DTPA extractant( Soil : solution in ratio of 1:2) and shake for 2 hour on a mechanical shaker.
3. Filter the solution through Whatman No. 42 filter paper
4. Feed the standard working solutions and prepare a standard curve by plotting AAS reading against the element concentrations.
5. Run a blank solution containing all reagents except soil.

*Precautions*

(1) The water to be used should always be double glass distilled water.

(2) The chemicals to be used must be extra pure or analytical grade reagent.

(3) The acetylene cylinders should be run only when the pressure is higher than 500 kPa (72.5 PSi). The acetylene lines should not be operated when the pressure is above 100 kPa (14.5 PSi), because at such higher pressure acetylene can spontaneously decompose or explode.

(4) If air flow is below 6 flow unit, the flame should not be ignited.

(5) The acetylene cylinder must always be in the vertical position so that the liquid acetone do not enter in the gas line.

(6) The nitrous oxide-acetylene flame should always be run with red feather visible or with less than 5 flow units of acetylene.

(7) Containers of the volatile organic solvents should not be left uncovered near the uncovered flame.

(8) The flame should not be left unattended.

(9) Without the aid of safety glasses or the flame shield, none should look at the flame.

(10) After calibration of the instrument, the air (or $N_2O$) and gas supply should not be adjusted to alter the sensitivity of the instrument.

***Calculations***

Available (DTPA extractable) micronutrient in soil (mg/ kg) = A x 2Where, A is the concentration of the element obtained from AAS.

## 9. Determination of Hot Water Soluble Boron (Berger and Truog, 1939) and Colorimetric Estimation Using Azomethine-H (John *et al.*, 1975)

***Principle***

Boron in soil exists both in organic and inorganic forms. Boron in soils is primarily in the +3 oxidation state and taken up by the plants in the form of borate anion. Boron deficiencies can be found most often in humid regions or in sandy soils. Boron is subject to loss by leaching, particularly in sandy soils and thus, responses to B are common in sandy soils. Responses to B have been found on a variety of crops in many countries. In contrast, B toxicity can be found mostly in arid and semiarid regions either due to high B in soils or high B containing irrigation water. Soil extraction with hot water continues to be the most acceptable method for predicting the availability of B in soils. Boron can be estimated colorimetrically using Azomethine-H as colour developing reagent.

This technique is rapid, reliable and more convenient to use than traditional procedures employing carmine or curcumin.

***Materials required***

1. Distillation flask and condensers
2. Reagent bottle
3. Funnel
4. Filter paper
5. Pipette
6. Measuring cylinder
7. Volumetric flask (25 ml, 100ml and 1000 ml)
8. Spectrophotometer

*Reagents*

1. 0.01 M $CaCl_2$ : Dissolve 1.47 g of AR grade $CaCl_2 .2H_2O$ in 200 ml distilled water and make the volume to one litre .
2. Buffer solution: Dissolve 250 g of ammonium acetate and 15 g of EDTA (Ethylene diamine tetra acetic acid) in 400 ml of distilled water and add slowly 125 ml of glacial acetic acid.
3. 1% L-Ascorbic acid solution: Dissolve 1 g of L-Ascorbic acid in 100 ml of distilled water.
4. Azomethine-H reagent: Dissolve 0.45 g of azomethine –H in 100 ml of 1 % L- ascorbic acid solution. Prepare a fresh reagent, when it is needed. Store in polythene bottle in a refrigerator

*Procedure*

1. Weigh 20 g of soil in a distillation flask and add 40 ml of 0.01M$CaCl_2$ solution.
2. Attach a water cooled reflux condenser
3. Heat the flask until first sign of boiling and reflux the contents for exactly 5 minutes
4. Allow to cool without removing the condenser
5. After cooling, filter the solution in a reagent bottle
6. Transfer 15 ml of filtrate into a 25 ml volumetric flask

7. Add 2 ml of buffer solution and 2 ml of Azomethine-H solution
8. Mix thoroughly and after 30 minutes read the absorbance at 420nm.
9. Run a blank (without soil) also by following the above steps
10. Derive the concentration of B in soil extracts from the standard curve.

**Preparation of Standard Curve**

1. Boron standard stock solution: Dissolve 0.570 g boric acid in one litre of distilled water, to obtain a stock solution of 100 ppm.
2. Working standard B solution: Take 5 ml of the stock solution (100ppm) in a 100 ml volumetric flask and dilute to 100 ml, this will give 5 ppm B solution.
3. Take 0, 0.5, 1.0, 1.5 and 2.0 ml of 5 ppm B solution into five numbered 25 ml volumetric flask to get 0, 0.1, 0.2, 0.3 and 0.4 ppm B solution
4. Add 2 ml buffer solution and 2 ml Azomethine-H solution and after half an hour take the reading at 420nm on a spectrophotometer.

*Calculation*

$$\text{Available Boron (ppm)} = \frac{\text{A X ml of } CaCl_2 \text{ added to soil}}{\text{Weight of soil (g)}} \times \frac{\text{Final volume}}{\text{vol.of filtrate (ml)}}$$

Where, A is concentration of B (ppm) obtained from graph

Available Boron Kg/ha = Available boron (ppm) x 2.24

## 10. Determination of Available Molybdenum in Soil

Molybdenum in soils is primarily in the+6 oxidation state, taken up by the plants in the form of molybdate anion, $MoO_4^{2-}$. Studies on the extraction of available Mo from soils have been limited. Further, the extremely low amounts of available Mo in soils under deficiency conditions make it difficult to determine Mo accurately. The accumulation of Mo in plants mostly is not related to total concentrations of Mo in soils, but rather to available Mo in soils. A variety of extractants have been used in attempts to extract available Mo in soils, although no routine soil test for Mo is available. Molybdenum deficiencies are rare and are mostly of concern for leguminous crops and acid soils($pH<5.0$). Since excessive Mo in forages can harm animal health, Mo fertilization is usually based on visual deficiency symptoms and/or history of crop rotation. Many extractants have been employed for the assessment of available Mo in soils. These extractants are: ammonium oxalate, pH 3.3; water; hot water, anion-exchange resin; AB-DTPA; ammonium carbonate; and Fe oxide strips. The most commonly used extractant for assessing Mo availability in soils has been ammonium oxalate (Grigg's reagent).

***Apparatus***

i. Rack for separating funnels

ii. Orbital shaker

iii. Spectrophotometer

*Reagents*

1. Ammonium oxalate (0.2 M buffered to pH 3.0): Dissolve 24.9 g of ammonium oxalate and 12.605 g of oxalic acid in approximately 800 ml of double distilled water and make to the mark in a one litre volumetric flask.
2. Stannous chloride : Dissolve 20 g of stannous chloride in 20 ml of Conc. HCl, make to 200 ml with distilled water in a 200 ml volumetric flask.
3. Ferric chloride: Dissolve 49 g of ferric chloride hexahydrate in distilled water and make to 1 litre in a volumetric flask.
4. Sodium nitrate: Dissolve 42.5 g of $NaNO_3$ in distilled water and make to 100ml in a volumetric flask.
5. Ammonium thiocyanate ($NH_4SCN$): Dissolve 50 g of $NH_4SCN$ in distilled water and make to 500ml in a volumetric flask.
6. Isopropyl ether: This ether is washed before use with a mixture of $SnCl_2$:$NH_4SCN$: distilled water in the ratio of 1:1:1. One-tenth volume of wash solution is used.

*Procedure*

1. Add 25 g air-dried soil, screened through a 2 mm sieve, to a 250 ml conical flask.
2. Add 150 ml of the buffered (pH 3.0) 0.2 M ammonium oxalate solution and shake for 16 h at room temperature, using an orbital shaker at 200 rpm.
3. Filter the extraction through Whatman No. 42 filter paper or equivalent. Centri-fuge the filtrate for 20 min.
4. Standard Mo solution (100 ppm Mo): Dissolve 0.15g of AR grade Molybenum trioxide ($MoO_3$) in 10 ml of 0.1NNaOH, make slightly

acid i.e. with HCl and make to 1 litre with distilled water in a volumetric flask.

5. Working standard solution (1.0 ppm Mo): Dilute 1 ml of the standard solution with distilled water to 100ml in a volumetric flask.
6. Take 0, 0.5, 1.0, 1.5 and 2.0 ml of 1.0 ppm Mo solution into five numbered separatory funnels and the final volume 25 ml to get 0, 0.02, 0.04, 0.06 and 0.08 ppm Mo solution.
7. Add 5 ml of 6N HCl to a separatory funnel
8. Transfer 5 ml of filtrate into a separatory funnel
9. Add in sequence 1ml of $FeCl_3$, 1ml of $NaNO_3$, 5 ml of $NH_4SCN$, and 4 ml of $SnCl_2$.
10. After addition of each solution shake, remove and rinse the ground glass stopper and the neck of separatory funnel with distilled water using a fine tipped wash bottle.
11. Add 5 ml of washed isopropyl ether and shake.
12. Allow the ether phase to separate, remove the ground glass stopper, rinse and wipe off the neck of funnel with tissue and draw off the aqueous phase.
13. Transfer the ether phase to a 15 ml centrifuge tube by pouring from the top of the separatory funnel.
14. Centrifuge for 5 minutes.
15. Transfer to a cuvette and measure the absorbance at 475nm in a spectrophotometer.
16. Run a blank (without soil) also by following the above steps.
17. Derive the concentration of Mo in soil extracts from the standard curve.

Available Molybdenum (ppm) = A x 30

Where, A is concentration of Mo (ppm) obtained from graph.

## 11. Estimation of Lime Requirement by SMP buffer method

The amount of lime or liming material required, depends on the nature of the soil and the magnitude of acidity. Through liming, the exchangeable $H^+$ is neutralized and the $CaCO_3$ equivalence of the exchangeable $H^+$ of soil is called 'lime requirement' (LR). By way of replacement of $H^+$ and $Al^{3+}$ ions, the $Ca^{2+}$ ions raised the percent base saturation, which leads to a corresponding rise in soil pH. There are various methods for the determination of LR in soils, of which, Shoemaker *et al.* (1961) is widely used.

**Equipments:** 1. pH meter, Volumetric flasks of various volume, Beaker, Balance,Glass rod etc.

*Reagents*

1. Extractant pH buffer solution

Dissolve 1.8 g of p-nitrophenol, 2.5 ml of triethanol amine (TEA), 3.0 g potassium chromate, 2.0 g of calcium acetate and 53.1 g of $CaCl_2.2H_2O$ in a 1000 ml of distilled water adjusting pH of the solution to 7.5 by dilute NaOH solution.

2. pH buffer solution: pH 4.0, 7.0 and 9.2 solutions or tablets are to be used for the calibration of pH meter.

*Procedure*

1. Take 10 g of soil in a beaker.
2. Add 10 ml of water and stir the contents.
3. Add 20 ml of SMP buffer to the soil slurry i.e. twice the volume of water used.
4. Shake the soil-water-buffer solution on an end-to-end shaker for 10 minutes and let set for 30 minutes after shaking. Alternatively, shake for 15 minutes and let set for 15 minutes after shaking. If using an automated pH analyser, stir the buffer with soil vigorously and allow solution to set for at least 30 minutes.
5. Calibrate pH meter and electrode using pH 4 and 7 buffers.
6. Place electrode in the soil slurry to measure pH. Measurement may be taken with or without continuous stirring. If measurement is made without continuous stirring, stir the sample before placing electrode in the sample.
7. Allow adequate time for pH to reach a stable reading.

***Calculation***

After determining the pH of the soil-buffer suspension, find out the amount of lime required to bring the soil pH to a desired level (6.4 and 6.8) as supplied in table for lime requirement.

**Table :** Lime requirement of soil

| Measured pH of soil- buffer suspension | Tons of pure $CaCO_3$ per acre required to bring change in pH | |
|---|---|---|
| | pH 6.4 | pH 6.8 |
| 4.8 | 10.6 | 12.4 |
| 4.9 | 10.1 | 11.8 |
| 5.0 | 9.6 | 11.2 |
| 5.0 | 9.1 | 10.6 |
| 5.2 | 8.6 | 10.0 |
| 5.3 | 8.2 | 9.4 |
| 5.4 | 7.7 | 8.9 |
| 5.5 | 7.2 | 8.3 |
| 5.6 | 6.7 | 7.1 |
| 5.7 | 6.2 | 7.1 |
| 5.8 | 5.7 | 6.5 |
| 5.9 | 5.2 | 6.0 |
| 6.0 | 4.7 | 5.4 |
| 6.1 | 4.2 | 4.8 |
| 6.2 | 3.7 | 4.2 |
| 6.3 | 3.2 | 3.7 |
| 6.4 | 2.7 | 3.1 |
| 6.5 | 2.2 | 2.5 |
| 6.6 | 1.7 | 1.9 |
| 6.7 | 1.2 | 1.4 |

*Precaution*

The p-nitrophenol and potassium chromate in the SMP buffer are hazardous and proper disposal of these hazardous materials should be followed.

## 12. Determination of Gypsum Requirement of Alkali Soils

Gypsum requirement (GR) of a sodic soil can be determined by treating the soil with known amount of excess saturated gypsum solution ($CaSO_4$, $2H_2O$), and then estimating the unutilized amount by Versenate (EDTA) titration method as suggested by Schoonover (1952). However, there are some other methods, for its determination, but this method is more suitable. Soils with high pH (8.5) may be reclaimed by treating them with a suitable amendment like gypsum, pyrite, saw dust, sulphur, etc. The reclamation is based on the principle of replacement for $Na^+$ by $Ca^{2+}$ ions from the adsorption sites.

***Equipments***

1. Volumetric flask (1000 ml)
2. Conical flask with rubber cork (100 ml and 250 ml)
3. Pipette (10 ml and 5 ml)
4. Measuring cylinder (1000 ml and 500 ml)
5. Balance
6. Mechanical shaker

*Reagents*

1. **Standard Gypsum solution:** Dissolve 5 g chemically pure $CaSO_4.2H_2O$ in 1 litre of distilled water and then filter with Whatman No. 1 filter paper.
2. **0.01 N $CaCl_2$ solution:** Dissolve 0.5 g AR/GR grade $CaCO_3$ powder in about 10 ml of 1 : 3 diluted HCl and make up the volume to 1 litre in a volumetric flask.
3. **0.01 N Versenate solution:** Dissolve 2.0 g pure EDTA disodium salt and 0.05 g of $MgCl_2$ (AR/GR grade) in distilled water and make up the volume to 1 litre in a volumetric flask.
4. **Eriochrome black T indicator (EBT):** Dissolve 0.5 g EBT and 4.5 g of hydroxylamine hydrochloride in 100 ml of 95% ethanol.
5. **Ammonium hydroxide-ammonium chloride buffer solution:** Dissolve 67.5 g of pure $NH_4Cl$ in about 570 ml of concentrated ammonia solution and dilute to 1 litre by adjusting pH of the solution to 10.0 with either dilute HCl or dilute $NH_4OH$.

*Procedure*

1. Take 5 g finely powered soil in a 250 ml conical flask
2. Add 100 ml of saturated gypsum solution
3. Shake in a mechanical shaker for 5 minutes
4. Filter through What No. 1 filter paper
5. Take a suitable amount of aliqout (5ml) in a china clay dish and dilute to around 100ml.
6. Adjust to pH 10.0 by adding 5ml of $NH_4Cl$-$NH_4OH$ buffer.

7. Add 10 drops each of masking agents (KCN, potassium ferrocyanide and hydroxyl amine) and 10 drops of EBT indicator.
8. Add 1 ml of $NH_4OH.NH_4Cl$ buffer solution +2-3 drops of EBT indicator
9. Place the dish on a magnetic stirrer plate and commence stirring, while titrating the solution with 0.01N EDTA solution under fluorescent light till the colour changes from red to bright blue.
10. Carry out a blank in exactly the same manner, taking 5ml of saturated gypsum solution instead of aliquot containing Ca.
11. Deduct the blank reading from the sample reading.

*Calculations*

Ca or Ca + Mg (me/l) in the aliquot = 2 V,

where 'V' stands for volume of versenate solution.

Since 1 litre extract = 50 g soil (5 g soil to 100 ml)

Ca retained or Ca requirement (mg/100 g) = (2V for added gypsum solution – 2 V for filtrate) × 2……(A)

Gypsum requirement (GR) of soil in t/ha furrow slice (15 cm) = (A)× 2

Correction factor may be applied depending on the purity of gypsum.

## References

Bray, R.H. and Kurtz, L.T. (1945) Determination of total, organic and available forms of Phosphorus in soils. Soil Sci. 59:39-45.

Berger, K.C. and Troug, E. (1961) Determination of Mo in biological material by dithiol-control of copper interference. J. Agric. Food Chem. 11:130-131.

Bremner, J.M. and Keeney, D.R. (1965) Steam distillation methods for determination of ammonium, nitrate and nitrite. Analytica Chemica Acta 32:485-495.

Bremner, J.M. and Mulvaney, C.S. (1982) Nitrogen total. In Page, AL(ed.) Methods of Soil Analysis. Agron. No.9 Part 2 : Chemical and Microbiological Properties, 2nd edition. Amer. Soc. Agron. Madison, WI, USA, pp. 595-624.

Buresh, R, J, Austin, E.R. and Crasell, E.T. (1982) Analytical Methods in N-15 Research. Fert. Res. 3 : 37-62.

Datta, N.P., Khera, M.S. and Saini, T.R. (1962) A rapid Colorimetric Procedure for the determination of Organic Carbon in Soils. J. Indian Soc. Soil Sci.10:67-74.

Hanway, J.J. and Heidel, H. (1952) Soil analysis methods as used in Iowa State College Soil Tesing Laboratory. Iowa Agric.57:1-31.

John, M.K. etal. (1975) Application of improved azomethane-H method to the determination of boron in soils and plants. Anal. Lett. 8:559-568.

Keeney, D.R. and Nelson, D.W. (1982) Nitrogen-inorganic forms. In Page *et al.*(eds). In Methods of Soil Analysis, Part 2, Agronomy Monograph No.9, Am.Soc.Agron.WI, USA.643-698.

Lindsay, W.L. and Norwell, W.A. (1978) Development of DTPA soil test for Zinc, Iron, Manganese and Copper. Soil Sci. Soc. Am. J. 42:421-428.

Olsen, S.R.etal. (1954) Estimation of available phosphorus in soils by extraction with sodium bicarbonate. Circular No. 939. U.S. Dep. Agric. Washington, DC, USA.

Scoemaker, H.E., McLean, E.O., and Pratt, P.F. (1961) Buffer methods for determining lime requirement of soils with appreciable amounts of extractable aluminium.Proc.Soil Sci. Soc. Am. 25:398-401.

Schoonover,W.R. (1952). Examination of Soils for alkali.ext.Bull.Univ. of California Extension Services, Berkeley, California (Mimeographed Publication).

Subbiah, B.V. and Asija, G.L. (1956) Arapid procedure for the determination of available nitrogen in soils. Curr. Sci. 25:259-260.

Walkley, A. and Black, C.A. (1934) An examination of different methods for determining soil organic matter and a proposed modification of the chromic acid titration method. Soil Sci. 37:29-38.

Williams, C.H. and Steinbergs, A (1969) Soil sulphur fractions as chemical indices of available sulphur in some Australian Soils. Aust. J. Agric. Res. 10 : 340-352.

❑❑❑

# D. Setting up Soil Testing Laboratory

## Historical background of Soil Testing Service in India

Soil test provides a means for assessing the fertility status of a soil, but soil test does not provide a direct measure of the total quantity of plant nutrient available in soil. Indeed, soil testing is a chemical method of estimation of plant available nutrients based on the principle of 3 E's (extractant, equilibration and estimation). The measured quantity of extractable nutrient is used to predict the crop response to applied nutrients be it fertilizers, manures or amendments and to obtain higher economic yield with minimum hazard to natural resources. The soil testing programme was started in India during 1955-56 with the setting up of 16 soil testing laboratories under the Indo-US Operational Agreement for "Determination of Soil Fertility and Fertilizer Use". Later, the State Governments and the Fertilizer industry started setting-up soil testing laboratories. In addition to 700 soil testing laboratories in the country now, under the National Project on Management of Soil Health and Fertility during XI Plan proposed for establishment of around 750 soil testing laboratories. When these laboratories become fully functional total sample analysing capacity will become 14.5 million per year by assuming the total analysing capacity of a STL @ 10000 samples/yr/lab. However, still there will be a big gap between the projected capacity and the number of farm holding, which is over 115 million in the country, that ideally require sampling, analysis and fertilizer use recommendation annually and which if not possible, at least after a gap of three years period as a practical measure. Capacity of the soil testing laboratories to analyse soil samples has always been inadequate and it continues to be so. As per the latest information,there are 1087(930 static and 157 mobile) STLs are functional in the country. Soil and water testing facilities have been created in 390 KVKs and 32 agri-clinics/agri-business centres.The number of STLs is likely to increase with time especially in the KVKs and Zonal Research Centres.

SWOT analysis on the soil testing service in India, indicate the following:

*Strength*

1. Soil test reports generally provide with appropriate fertilizer application recommendations for nitrogen, phosphorous, potassium and lime. Thus, soil tests assess the amount of nutrients already present and add the balance amount of nutrients required for a desired production level.
2. One of the most important basic information from soil test is revealed is, soil pH. Most of the Indian farmers are not aware about the acidity (low pH), or alkalinity (high pH) of their fields. The acidity in the soil

limits plant growth to such an extent that plants will be unable to utilize the applied fertilizers. Acid soils also make phosphorus unavailable to plants. On the other end, a high pH can cause micronutrients to become unavailable to plants. Thus, soil pH can be adjusted to an optimum level with the help of soil test report.

3. In India, where most of the farmers are poor, by reducing fertilizer application or by applying only what is needed to get the same crop yield, they can save significant amount of money and time spent on purchasing extra quantity of chemical fertilizers.
4. Soil test based fertilizer recommendation would lead to reduce such kind of soil/water contamination and help maintain and sustain soil fertility and reduce risk of crop damage due to nutrient deficiency/ toxicity.
5. Soil testing also allows for determining the secondary and micronutrient requirements of crops. Thus, proper balance of plant nutrients can be maintained.

*Weakness*

1. Soil test results represent a portion of the total soil nutrients, that is extracted from the soil by chemical solutions as analysis is usually confined to the surface soil. No information is collected on the fertility status of lower depths in the root zone.
2. Soil test laboratories often use different extracting chemicals and provide soil test values. It is therefore, more appropriate to look at the interpretation of soil test results (sufficient, deficient, high medium, low, etc.) than to look at the numbers.
3. Soil testing is only fair, at best, for detecting crop response to micronutrients. Micronutrients occur in very small concentrations in the soil, and field trial data correlating crop response to soil test levels of micronutrients are lacking. If a micronutrient deficiency is suspected, it is usually better to analyse the plant tissue rather than the soil.
4. More often, fertilizer requirements for crops are recommended based on soil test values on the basis of research experiments and follow up trials, where irrigation water is not a limiting factor. But, in reality, most of the resource poor farmers, who don't have reliable source of irrigation can't give required number of irrigation. Therefore, the target yield falls short of the expectation.
5. In absence of Soil Testing Laboratories (STLs) in nearby area, it is

very difficult for each farmer with smaller/fragmented holding to get their soils tested. Farmers are also not aware of soil sampling in proper manner and the authenticity of soil test results highly depend on the accuracy of soil sampling.

*Opportunities*

1. Soil testing is recommended to be done each year after crop harvest, to know the available nutrient status. However, in India, most of the farmers are unaware of the advantages of this service.Therefore, farmers need to be educated with proper care in extending this facility to farmers.
2. There is a need to reorient the research strategies in the field of soil testing methodologies, so as to make it simple and understandable by all stakeholders. Uniform methodologies for analytical purposes by all the STLs, will increase the reliability of the reports .
3. Recommendation of fertilizer requirement should be done keeping in view the resource status of the farmers. This requires change in research strategies to standardize nutrient requirements under various scenarios of irrigation availability, financial condition of the farmers, agro-climatic conditions, etc.
4. The analytical procedures should be sensitive, rapid, adaptable for routine use, and cost-effective for its clients. These essential requirements underpin modern soil testing, and, where varied, seriously undermine the reliability of advice given to clients and the credibility of soil testing itself. In addition, unreasonable service charges hinder wider adoption of soil testing.
5. Overall, there exists huge opportunities for soil testing services in India to expand and make the fertilizer application in a more efficient manner. This would not only reduce the burden on national exchequer, which are spent on providing fertilizer subsidy, but also help the farmers in realizing betterprofits.Blanket recommendation of fertilisers for crops need to be avoided.

*Threats*

1. Soil testing itself doesn't have any threat, as it is a pre-requisite for efficient farm management, if weaknesses, as mentioned above are eliminated.
2. Fertilizer recommendation needs to be done on the basis of soil test

values as well as socio-economic and other farm conditions of individual farmer. Soil tests without farmers active participation will not lead to success of the programme.

3. Rapid and cost-effective reporting is necessary for its wider adoption among all the farmers.
4. Interpretation of soil test results is often poor.Farmers often get results very late in the crop season, causing disinterest among them.

***Requirements for establishment of Soil Testing Laboratory***

1. Building
2. Technical personnel
3. Instruments
4. Analytical techniques and protocol
5. Data station and softwares
6. Laboratory information and management system
7. Communication facility

In order to establish a soil testing laboratory, the building should have space for soil sample storage, sample preparation, storage of chemicals, working table to carry out the analysis, instrumentation room and computer room well equipped with internet facilities for communication. Technical personnels having basic knowledge of chemistry/agriculture atleast three in number, field assistants two in number for soil sample collection and processing and over all in charge of the laboratory having M.Sc. (Ag) degree to supervise analytical work, interpretation of soil test results, conducting field experiments and refinement in soil testing methodology are required for establishing a small to medium class STL. The instruments required for a soil testing laboratory are pH meter, conductivity bridge, colorimeter or spectrophotometer, flame photometer, glass assembled digestion & distillation for N analysis, distillation unit for preparation of distilled water, to and fro shaker or orbital shaker, physical balance to weigh soil samples and a precision balance for weighing of chemicals. In addition to the instruments glassware, chemicals and other petty items are to be kept in a soil testing laboratory.

In soil testing laboratories, the use of acids, alkalis and some hazardous and explosive chemicals is inescapable. Apart from this, some chemical reactions during the process of analysis, may release toxic gases and if not handled well, may cause explosion. Inflammable gases are also used as a fuel/heating source. Thus, safe working in a chemical laboratory needs special care, both in terms of design and construction of the laboratory building, and handling and use of chemicals. For chemical operations, special chambers also need to be provided. Air temperature of the laboratory and working rooms should be maintained at a constant level between 20-25$^0$ C. Humidity should be kept at about 50%, since soil samples are affected by the temperature and humidity. Even some chemical operations get influenced by the temperature. Hence, maintenance of temperature and humidity as specified is critical.Proper air circulation is also important so as, hazardous and toxic fumes and gases do not stay in the laboratory for long. The release of gases and fumes in some specific analytical operation are controlled through fume hood or trapped in acidic/alkaline solutions and washed through flowing water. Maintenance of clean and hygienic environment in the laboratory, is essential for the good health of the workers.

Care should be taken to store acids and hazardous chemicals in separate and safe racks. An inventory of all the equipment, chemicals, glassware and miscellaneous items in a laboratory should be maintained.

Room 1. Reception/sample receipt/dispatch of reports.

Room 2. Sample storage and preparation room.

Room 3. Nitrogen digestion/distillation room (with fume hood for digestion).

Room 4. Instrument room

Room 5. Sample processing room :

- To prepare reagents, chemicals and to carry out their standardization.
- To carry out extraction of soil samples with appropriate chemicals/ reagents.
- To carry out titration, colour development, filtration, etc.
- All other types of chemical work.

Room 6. Storage room for chemicals and spare equipments.

Room 7. Mechanical analysis room

Room 8. Conference / training/ meeting room

**Layout plan of a soil testing laboratory**

| | 6 m | 2 m | 4 m | 3 m | |
|---|---|---|---|---|---|
| 5 m | Instrument room AAS, UV-VIS, pH meter, EC-TDS | | Balance, Oven, Refrigerator, Test kits, centrifuge | Store room | 5 m |
| 4 m | Soil sample processing room equipped with fume hood | | Physical balance, shakers, working table for preparation of reagents | Soil sample storage room | 7 m |
| 5 m | Water Distillation plant, Kjeldahl digestion & distillation room fitted with fume hood, Generator | | Chamber of officer I/c of STL | Conference hall | 4 m |
| 5 m | Training hall | Gate | Reception | Office Computer | 3 m |

***Soil test Interpretation***

The basic concepts in interpreting soil test values and in making fertilizer recommendation have emerged since 1950's. Two of these follow the general concept that, there are definable levels of individual nutrients in the soil below which, crops respond to fertilizers with some probability and above which they are not likely to respond. One of these approaches is to, increase the soil test value up to the level at which no yield increase occurs and then maintain it by applying an amount equal to crop removal. This is build up and maintenance approach. Second approach is termed sufficiency concept. This is based on the establishment of low, medium and high category of available nutrients in soil. The third approach is termed as basic cation saturation ration (BCSR) concept,which emphasizes on the concept of maximum yields,that can be achieved by creating an ideal ratio of calcium, magnesium and potassium in soil. In general, most nutrient recommendations of soil testing laboratories use the sufficiency concept.

***Soil fertility as a nutrient index***

Generally, the soil testing laboratories use organic carbon as an index of available N, Olsen's or Bray's method for available P and neutral normal ammonium acetate for available K. In semi-arid tropics, nitrate nitrogen is also used as an index of available N in soil. This approach is too general and may be only indicative for National level appreciation of soil fertility status and not for the benefit of individual farmers. This classification needs refinement.

Available nutrient status in the soils is generally classified as low, medium and high, which are generally followed at the National level are given below.For different soil types/agro-climatic zones,critical levels of nutrients need to be delineated and rating charts need to be established and used for interpretation of soil test results.

**Rating limits for soil test values used in India**

| Sl. No. | Soil parameters | Soil fertility status Low | | |
|---|---|---|---|---|
| | | Low | Medium | High |
| 1. | Organic carbon (%) | <0.5 | 0.5-0.75 | >0.75 |
| 2. | Available N-$KMnO_4$ (kg N /ha) | <280 | 280-560 | >560 |
| 3. | Available P (kg /ha) | <10 | 10-25 | >25 |
| 4. | Available K (kg /ha) | <108 | 108-280 | >280 |

Nutrient index is calculated based on the nutrient status of the area and used to indicate fertility status of soils for the purpose of mapping. The utility of preparing soil fertility maps continues to be recognized in view of the fact that such maps can be used for general planning of fertilizer supply / use on area basis, in the absence of a specific soil analysis for each and every farm holding.

The following equation is used to calculate Nutrient Index Value:-

$$\text{Nutrient Index} = \frac{(Nl \times 1) + (Nm \times 2) + (Nh \times 3)}{Nt}$$

Nt = Total number of samples analysed for a nutrient in any given area. Nl = Number of samples falling in low category of nutrient status. Nm = Number of samples falling in medium category of nutrient status. Nh = Number of samples falling in high category of nutrient status.

Separate indices are calculated for different nutrients like nitrogen, phosphorus and potassium. Parker had classified the nutrient index values less than 1.5, as the indicative of low nutrient status and between 1.5 to 2.5, as medium while higher than 2.5, as high nutrient status. Ramamoorthy and Bajaj had categorised these values as less than 1.7, being indicative of low fertility status, between 1.71 to 2.33, as medium and more than 2.33, to classify as high.

Secondary and micronutrient deficiencies are becoming critical, with the intensification of agriculture and use of high analysis chemical fertilizers. Thus, micro nutrient testing facilities are also required to be created in the soil testing labs. Under Indo-UK bilateral programme on strengthening the soil testing facilities in India during 1980, Atomic absorption spectrophotometers were supplied to the soil testing labs in 20 states. Apart from this, some state governments have also provided this equipment to the soil testing labs under the state plans. Government of India is strengthening the soil testing labs by providing funds for this equipment. Thus, the testing of micro-nutrients was started by the soil testing labs. State Agricultural Universities and coordinated ICAR scheme on micro nutrient in soils and Plants are delineating micro nutrient deficient areas and setting standards for micronutrient sufficiency and deficiency in soils and plants. Delineation work on secondary and micronutrients established the critical levels of these nutrients below which, there is a chance of getting response of applied fertilizers. Critical level of secondary and micronutrients in soil, based on such studies are given below.

| Nutrients | Deficient | Sufficient |
|---|---|---|
| Exchangeable Ca (Cmol ($P^+$) $kg^{-1}$) | Below 1.5 | Above 1.5 |
| Exchangeable Mg (Cmol ($P^+$) $kg^{-1}$) | Below 0.5 | Above 0.5 |
| Available S 0.15% $Cacl_2$ (mg $kg^{-1}$) | Below 10 | Above 10 |
| Available Zn DTPA extractable (mg $kg^{-1}$) | Below 0.5 | Above 0.5 |
| Available Cu DTPA extractable (mg $kg^{-1}$) | Below 0.2 | Above 0.2 |
| Available Fe DTPA extractable (mg $kg^{-1}$) | Below 4.5 | Above 4.5 |
| Available Mn DTPA extractable (mg $kg^{-1}$) | Below 2.0 | Above 2.0 |
| Available B Hot water extractable(mg $kg^{-1}$) | Below 0.5 | Above 0.5 |
| Available Mo Ammonium oxalate extractable (mg $kg^{-1}$) | Below 0.2 | Above 0.2 |

***Use of test kits in soil testing laboratories***

At present, there are about 1200 soil test laboratories in India, which cannot cater to the needs of soil testing of the farming sector. Owing to problems of easy access of many laboratories to farmers, distance, transpotation costs, and lack of awareness, many farmers do not get soil testing done at regular intervals. With increase in soil problems, such as acidity, erosion, salinity, sodicity and emerging micronutrient and secondary nutrient deficiencies, the need of soil testing cannot be ignored. Further, soil health concerns, poor response of added nutrients in many crops, quality of produce and adverse impact of high rate of chemical fertilisers on crop yields are of great concerns for the farm sector. In this regard, soil test kits can fill in the gap to a considerable extent, if used properly.

Using the portable soil testing kit developed by various organizations, one can analyse soil at field level and know the soil fertility status on the spot. The test kit consists of chemical reagents for various soil parameters like pH, EC, organic carbon, nitrogen, phosphorus, potassium, sulphur, free calcium carbonate, zinc, iron, and boron and different colour shades are provided in the kit for each test. It works well both in mildly acidic soils, saline or alkaline soils and calcareous soils. Soil test kit gives both qualitative and quantitative evaluation of soil fertility mainly suitable for areas, where soil sampling andtransport of soil samples is difficult.

In addition to soil test kits, portable plant tissue test kits can also be used for diagnosis of nutrient deficiencies, which uses fresh plant material (whole leaves, leaf blades, or petioles). Use of portable plant test kit is within the reach of Indian farming community, which enables them to diagnose the deficient nutrient and plan remedy. These simplified field tests for green plant tissue indicate, if the growing plants are receiving adequate amounts of available nutrients from the soil or not. All tests give qualitative results for the specific nutrients. By comparing test results from healthy and problem plants, it is possible to pinpoint deficiencies or excessive nutrient conditions. Cost of routine sampling and analysis that involves many samples, drying them properly, processing them and analysing in the laboratory employing sophisticated equipment might be too costly for many crop growers and unaffordable by Indian farmers and also difficult to reach a marginal farmer.

Portable water Test Kit can determine pH, Electrical conductivity, total dissolved solids, total hardness, alkalinity and salinity, apart from cations like calcium, & Magnesium and anions like carbonates & bi-carbonates, sulphate, chlorides, nitrates, phosphates and fluorides, which are important from water quality, salinity hazard, plant nutrition and environmental points of view.

Similarly, organic manure kit for evaluation of manurial value of organic manures and estimations, are to know the amount of organic carbon (organic matter) present in each lot of their product as well as water soluble major nutrients like nitrogen, phosphorous, potassium and sulphur on the spot without loss of time.

Soil Science Division of IARI, Pusa, New Delhi has recently developed a Digital Soil Test and Fertiliser Recommendation Kit (STFR).This consists of a digital STFR meter, a mini shaker, an accessories kit, glasswares kit and a reagent kit, costing about Rs. 30,000/- only. This can boost the soil testing work at farmers' doorsteps and help in promoting soil test based nutrient use in crops.

### *Quality Control and Assessment Procedures*

1. Good laboratory practices (e.g., housekeeping, storage of chemicals, laboratory techniques) and good management practices (e.g., calibration, maintenance of equipment) are integral parts of quality control. The laboratory is maintained in a clean and organized manner. All chemicals are dated on receipt and disposed of, when shelf life is exceeded.

2. Methods are documented and followed.
3. Specific conductance of distilled water' is routinely checked. Double distilled water is used for trace element analysis.
4. Dilute working standards are prepared daily.
5. Matrix match is important in calibration.
6. Glassware and plasticware are rinsed with tap water immediately after use. For most analyses, rinsing with tap water followed by distilled water is sufficient. For certain analyses, however, washing with dilute HCl followed by thorough rinsing with distilled water is required.
7. Glassware is stored in dust-free cabinets.
8. Care is exercised in sampling and sample handling. Sample integrity is ensured.
9. Samples are stored according to their analytical requirements.
10. Operation and service manuals for all instrumentation are strictly followed. Preventive maintenance is essential. For example, balances are checked and serviced annually by trained service personnel. Records of downtime and service on equipment are maintained to assist in projecting repair and replace-ment needs.All sophisticated equipments in a STL should be under the AMC(annual maintenance contract).
11. Calibration of the instruments, for example pH meters are calibrated against two buffers bracketing the expected pH of the samples. Atomic absorption spectro-photometers, inductively coupled plasma spectrometers and other such instruments are calibrated with standard solutions for every batch of samples.Standards are checked every 20-50 samples, and at the end of each batch. After standardization of the instrument, accepted deviation of analytical results must range within 0-4% of the true value.
12. All details of the analytical work (worksheets) are filed as permanent records.
13. Number of significant integers: Only the last figure reported should be in doubt.
14. Samples received for analysis are checked for acceptability (e.g., sample condition, appropriate documentation) and lab numbers are assigned and noted in a log book. The log book records, names of submitters, consecutive serial(lab) numbers, date on which samples were received, were analysed and name of analyst.

15. Method blanks are required to correct for contamination in reagents and other materials (e.g., filter paper, acids, water). Method blanks are run for each group of samples analysed. This involves repetition of the entire procedure without including the sample. Blanks containing the matrix of the calibration standards are analysed at the beginning of each batch, after every 20-50 samples, and at the end of each batch.

16. Duplicate **samples** are used to determine within-run precision. Duplicate means to repeat the whole procedure. If an analysis is repeated because the first result appears anomalous, this should not be considered a duplicate. For routine analysis, one duplicate sample is run for every 20 samples to monitor the precision or reproducibility of the method. All relative standard deviations calculated from duplicate sample analysis should be within acceptable limits (5-15%, depending upon parameter and concentration). No further samples are analysed unless duplicate results are acceptable. The total within-laboratory standard deviation includes between-run (between-batch) and within-run (within-batch) variations.

17. Internal (performance) audits are performed using "blind" check samples. These are samples of known composition, that are given to the analyst without his or her knowledge. Blind samples are intermingled with and indistinguish-able from actual samples to ensure that they do not receive special treatment. To ensure valid data, known reference materials are run with each batch of samples. If results are not acceptable, corrective measures are taken before performing analysis on actual samples. Also, if the results are questionable, the analysis is repeated on those samples. Reference materials include the following:

    **(i) Internal** reference materials: Samples collected, prepared, and analysed by analysts within the Analytical laboratory.

    **(ii) External** reference materials: Samples analysed by different laboratories. To ensure that laboratories produce credible data, it is important to participate in inter-laboratory comparison studies.

### *Error, Precision, Accuracy and Detection Limit*

Error is an important component of the analysis. In any analysis, when the quantity is measured with the greatest exactness that the instrument, method and observer are capable of, it is found that the results of successive determination differ among themselves to a greater or lesser extent. The average value is accepted as most probable. This may not always be true value. In some cases, the difference in the successive values may be small, in some cases it may be large, the reliability of the result depends upon the magnitude of this difference. There could be a number of factors responsible for this difference, which is also referred as 'error'. The error in absolute term is the difference between the observed or measured value and the true or most probable value of the quantity measured. The absolute-error is a measure of the accuracy of the measurement. The accuracy of a determination may, therefore, is defined as the proximity between it and the true or most probable value. The relative error is the absolute error divided by the true or most probable value. The error may be caused due to any deviation from the prescribed steps required to be taken in analysis. The purity of chemicals, their concentration/strength and the accuracy of the instruments and the skill of the technician are important factors.

In analysis, other important terms to be understood are precision and accuracy. Precision is defined as the concordance of a series of measurements of the same quantity. The mean deviation or the relative mean deviation is a measure of precision. In quantitative analysis, the precision of a measurement rarely exceeds 1 to 2 parts per thousand. Accuracy expresses the correctness of a measurement, while precision expresses the reproducibility of a measurement. Precision always accompanies accuracy, but a high degree of precision does not imply accuracy. In ensuring high accuracy in analysis, accurate preparation of reagents including their perfect standardization is critical. Not only this, even the purity of chemicals is important.

In the analysis for trace elements in soils, plants and fertilizers and for environmental monitoring, need arises to measure very low contents. Modern equipments are capable of such estimation. However, while selecting an instrument and the testing method for such purpose, it is important to have information about the lowest limits upto which it can be detected or determined with sufficient confidence. Such limits are called as detection limits or lower limits of detection.

### *General rules and requirements – DO's and Don'ts in a laboratory*

- Learn safety rules and use first aid kits. Keep the first aid kit handy at a conspicuous working place in the laboratory.
- Personal safety aids such as laboratory coat, hand protection gloves, safety glasses, face shield and proper footwear should be used, while working in the laboratory.
- Observe normal laboratory safety practice in connecting equipment with power supply, in handling chemicals and preparing solutions of reagents. All electrical work must be done by qualified personnel.

- Maintain instrument manual and log book for each equipment to avoid mishandling, accident and damage to equipment.
- Calibrate the equipment periodically.
- Carry out standardization of reagents daily before use.
- Always carry out a blank sample analysis with each batch.
- Ensure rinsing of pipette before use with the next solution.
- Do not return the liquid reagents back into the bottle after they are taken out for use.
- Do not put readily soluble substances directly into volumetric flask, but first transfer into a beaker, dissolve and then put in the flask.
- Store oxidizing chemicals like iodine and silver nitrate only in amber colour bottles.
- Keep the working tables/space clean. Clean up spillage immediately.
- Wash hands after handling toxic / hazardous chemical.
- Never suck the chemicals with mouth but use automatic pipetting device.
- Use forceps / tongs to remove containers from the hot plates/ovens/ furnaces.
- Do not use laboratory glassware for eating/drinking.
- Use fume hood while handling concentrated acids, bases and hazardous chemicals.
- Never open a centrifuge cover until the machine has stopped.
- Add acid to water and not water to acid while diluting the acid.
- Always put labels on bottles, vessels and wash bottles containing reagents, solutions, samples and water.
- Handle perchloric acid in fume hoods.
- Do not heat glasswares and inflammable chemicals directly on the flame.
- Read the labels of the bottles before opening them.
- Do not hold stopper between fingers, while pouring liquid from bottle, nor put it on the working table but on a clean watch glass.
- Instruments are very costly and their repair is very time consuming. Therefore, handle them carefully.

- Take the help of laboratory attendant for your requirements of chemicals, solutions and glassware and any other clarification.

***First aid***

In spite of these precautions if any mishap occurs during the conduct of analysis, adopt the following first aid safety measures:

***Burns***

*a. Cloth burns*

Apply tannic acid jelly, Burnol or butesin picrate ointment on slight burns caused by dry heat in which the skin is not broken.

*b. Skin burns due to acid*

First wash with water thoroughly, then with saturated solution of $NaHCO_3$ and finally with water. If acid burns are serious, apply any disinfectant after the washing treatment. Allow the skin dry and cover the burnt skin with burnol.

*c. Alkali on the skin*

The treatment is same as in the case of acid burns except that in second washing use 1% acetic acid instead of saturated solution of $NaHCO_3$.

*d. Organic substances on the skin*

Wash the skin thoroughly with rectified spirit, then with soap and warm water.

***Cuts***

Allow minor cuts to bleed for a few seconds. If the cut is due to glassware breakage then remove the glass particles. Apply a disinfectant and bandage. In serious cuts, consult the doctor at once after washing the part with disinfectant.

***Eye accident***

In all case consult the doctor at once. Meanwhile, first aid should be done.

*a. Acids in the eyes*

Wash the eye with 1% $NaHCO_3$ if the acid is dilute. In case of concentrated, it should first be washed with a large amount of water and then thoroughly with the bicarbonate solution.

*b. Caustic alkali in the eyes*

Proceed as for acid in the eyes, but wash with 1% boric acid solution instead of bicarbonate.

***Poisons***

Acids: If acid is sucked by the mouth, drink a lot of water, followed by lime water or milk of magnesia.

Intake of caustic alkalies:Drink a lot of water followed by vinegar, lemon or orange juice, solution of citric or lactic acid.

Other poisons: If salts of heavy metals have swallowed, give milk or white of an egg. If compounds of arsenic or mercury have been swallowed, give one teaspoonful of mustard, or one teaspoonful of common salt or zinc sulphate, in a tumbler or warm water. Rush immediately for medical help.

Harmful gases: The victim should at once be removed to the open air and the clothing of the neck should be loosened. Inhale $NH_3$vapours to counteract $Cl_2$ or $Br_2$ fumes, if inhaled in small amounts. Gargles with $NaHCO_3$ solution are also helpful.

***Fires***

Turn out all the gas burners and switch off all the electric hot plates in the vicinity.

To sum up,**the improvements desired are:**

*1. Check common bottlenecks*

- Sample delivery
- Completeness of information sheet
- Sample analysis
- Data entry
- Report generation
- Reaching the report to the end-users

*2. Report results of soil testing*

- Value addition step
- Opinion about the laboratory is developed

- Concise for easy comprehension
- Need to apply vs., how much to apply
- Client address and history
- Results with interpretation
- Recommendations with expected results

*3. Overcome problems of Analytical service*

- Keep qualified managers and personnels
- Make it economically viable
- Ensure credibility
- Add value to the service provided
- Respect views of end-users

*4. Enhance quality*

- Use validated and reliable analytical protocols
- Test the proficiency periodically
- Collaborative network and peer assessment
- Full proof and efficient advisory service

❑❑❑

## CHAPTER - 5

# Plant Analysis

**S. C. Kotur**

Plant analysis includes tests on fresh tissue in the field and analysis of total nutrients in plants in the laboratory. Plant analysis is based on the relationship between nutrients in a plant and nutrient availability in a soil. Plant analysis is not a direct method of soil fertility evaluation, but it can serve as a valuable supplement to soil testing in the task of soil fertility evaluation. Plant analysis is useful in confirming nutrient deficiencies, toxicities and imbalance and identifying hidden hunger. Plant analysis generally provides more current plant-based information than soil testing, entails more effort in terms of sampling, sample handling and analysis and, perhaps, expensive.

Unlike soil analysis, where the amounts of available nutrients are of importance, in plant analysis, the total content of nutrients is relevant. Knowledge of nutrient concentration in growing plants can serve as a tool for correcting any deficiency,which the plant may suffer from, if detected early. The nutrient content in a healthy crop is considered an effective indicator of the requirement that may be used as a standard for the crop performance. It is often used to evaluate the efficacy of a fertiliser management programme. The information will help plan nutrient application in subsequent crops on that field or to compute nutrient removals in relation to productivity and nutrient balance sheet. Plant analysis is done with the following objectives:

1. To identify deficiency symptoms and to determine nutrient shortages before they appear as symptoms
2. To aid in determining the nutrient supplying capacity of a soil, as a complement to soil testing
3. To aid in determining the effect of nutrient addition on the nutrient supply in the plant
4. To study the relationship between nutrient status of the plant and crop performance

5. To help in assessing nutrient uptake in cropping systems and calculate balance sheet of nutrients in soil under long term fertiliser experiments.

## Sufficiency/Optimal Range of Nutrients in Plants

Motsara and Roy (2008) have provided information on the sufficiency or optimal range of nutrients for plants (Table 1). Although the nutrient contents vary from one species to another, the composition of a particular species, is generally considered to be indicative under optimal nutrient supply and growth conditions. This can generally help in understanding and detecting the probable deficiency of nutrients.

**Table 1 :** General sufficiency or optimal range of nutrients in plants

| Nutrients | Sufficiency/optimal range |
|---|---|
| N | 2.0-5.0% |
| P | 0.2-0.5% |
| K | 1.0-5.0% |
| Ca | 0.1-1.0% |
| Mg | 0.1-0.4% |
| S | 0.1-0.3% |
| Zn | 20-100 µg/g |
| Fe | 50-250 µg/g |
| Mn | 20-300 µg/g |
| Cu | 5-20 µg/g |
| Mo | 0.1-0.5 µg/g |
| B | 10-100 µg/g |
| Cl | 0.2-2.0% |

## Sampling Plant Samples

Leaf sampling technique: It is necessary to collect the leaf or petiole sample, that is considered the index tissue, which bears the closest empirical relationship with crop yield. The commonly accepted physiological stage is the 'youngest or recently mature leaf', which has attained stability of both mobile and immobile nutrients. This is because, the younger leaves have distinctly higher contents of mobile nutrients, while the older or mature leaves are richer in immobile nutrients. It is impractical to sample leaves of different stages of growth in routine leaf analysis. Therefore, the youngest mature leaf, serves best both the kinds of nutrients in one go.

Common guidelines for leaf and petiole sampling :

- Recently matured leaf is selected.
- Composite sample is taken for analysis.
- Leaves that are fully exposed to sunlight are to be collected.
- Samples are collected prior to fertiliser application.
- Diseased or insect damaged plant parts are avoided.
- Plants growing within areas of unusual feature are avoided.
- Plants under water or temperature stress are not sampled.
- Sample contamination is to be prevented to the extent possible.

Some examples of the plant parts to be sampled in case of agricultural and horticultural crops are presented in Tables 2 and 3. Fig.1 shows the method of sampling of banana.

**Table 2 :** Typical plant parts suggested for sampling

| Crop | Part to be sampled, with age or growth stage |
|---|---|
| Wheat | flag-leaf, before head emergence |
| Rice | $3^{rd}$ leaf from apex, at tillering |
| Maize | ear-leaf before tasseling |
| Barlcy | flag-leaf at head emergence |
| Pulses | recently matured leaf at bloom initiation |
| Groundnut | recently matured leaflets at maximum tillering |
| Soybean | $3^{rd}$ leaf from top,2 months after planting |
| Cotton | petiole, $4^{th}$ leaf from apex, at initiation of flowering |
| Sugarcane | $3^{rd}$ leaf from tip of young shoots |
| Potato | most recent, fully developed leaf (half grown) |
| Tomato | leaves adjacent to inflorescence (mid-bloom) |
| Onion | top non-white portion (1/3 to ½ grown) |
| Beans | uppermost, fully developed leaves |
| Pea | leaflets from most recent, fully developed leaves, at first bloom stage |

**Table 3 :** Plant tissue sampling guidelines for fruit crops

| Fruit crop | Index tissue | Growth stage/time |
|---|---|---|
| Avocado | Leaf | Fully expanded youngest leaves ,about 5 month old |
| Banana | Leaf lamina | Centre 20 $cm^2$ of leaf lamina on both sides of midrib of 3$^{rd}$ fully open leaf at 16$^{th}$ to 20$^{th}$ leaf stage |
| Custard apple | Leaf | 5$^{th}$ leaf from apex 2 months after new flush grows |
| Fig | Leaf | Fully expanded leaves, mid-shoot of current growth |
| Grape | Petiole | Petiole from 5$^{th}$ leaf position, 40-45 days after bud differentiation |
| Citrus | Leaf | 3-5 month old leaf from new flush |
| Guava | leaf | 3$^{rd}$ pair of recently mature leaf at bloom stage |
| Mango | Leaf + petiole | 4-7 month-old leaf from middle of shoot |
| Papaya | Petiole | 6$^{th}$ petiole from apex, 6 months after planting |
| Passion fruit | Leaf | Mature leaf opposite to the first open flower |
| Pineapple | Basal portion | middle 1/3$^{rd}$ portion of white basal portion of 4$^{th}$ leaf from apex 4-6 months after planting |
| Pomegranate | Leaf | 8$^{th}$ pair of leaf from new growth |
| Sapota | Leaf | 10$^{th}$ leaf from new flush |
| Phalsa | Leaf | 4$^{th}$ leaf from apex at new growth |
| Ber | Leaf | 5$^{th}$ and 6$^{th}$ leaf from growing tip from primary or secondary shoots |

Ref. Reuter and Robinson,1986, Jones *et. al.,* 1991

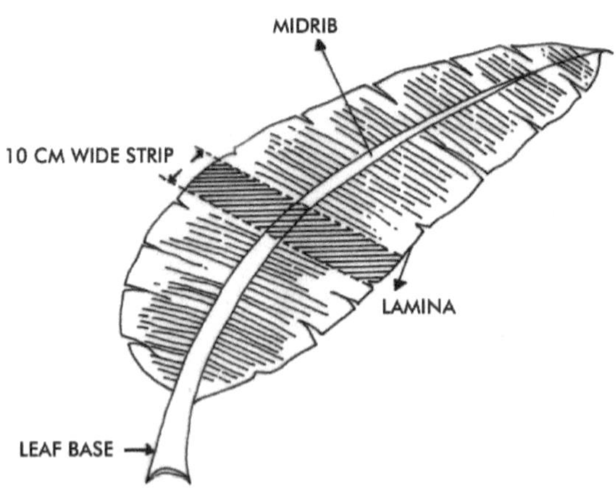


**Fig.1 :** Sampling of banana leaf sample

Based on the above guidelines, samples need to be collected. Some additional guidelines are discussed below :

- Select a vegetative terminal unless specified otherwise.
- Sample at 2m height or at chest level.
- Collect composite samples from all directions, north, south, east and west.
- To monitor leaf nutrient status on a regular basis, sample from the same part of the orchard.
- Select leaves fully exposed to sunshine avoiding the interior-liers.
- Collect leaves prior to irrigation, manure and fertilizer application.
- Collect about 50 white basal part of the 'D' leaf of pineapple; 30 medium leaves of ber or citrus; and 20 leaves of fruits, such as in mango, jackfruit or phalsa. In grape, 200 petioles are adequate. In banana, 10cm long lamina from one side of the 3rd fully open leaf from the randomly chosen 5-6 plants is sufficient.
- Exclude soiled, diseased, senescent or insect-affected leaves.
- Avoid sampling from plants located in unrepresentative or rocky spots.
- Avoid sampling, when the plants are under moisture or temperature stress.
- Sample during certain period of the day to avoid possible diurnal changes.
- Store the sample in fresh paper bags to prevent contamination before processing.

**Sample Handling & Preparation for Analysis**

- For analysis of seasonal crop plants,pick a few representative plants at random from each plot. Remove the shoot (aerial part) with the help of a sharp stainless steel cutter for whole shoot analysis or the desired part for analysis of specific plant parts.
- If roots are to be included, uproot the whole plant carefully from wet soil, retaining even the fine/active roots. Dip the plant roots gently in water several times to remove adhering soil.
- Wash with water several times.
- Wash the samples with about 0.2 % detergent solution to remove the waxy/greasy coating on the leaf surface.

- Wash with 0.1 *M*HCl followed by thorough washing with plenty of water. Give a final wash with distilled water.
- Wash with double distilled water, if micronutrient analysis is to be done.
- Soak to dry with tissue paper.
- Air-dry the samples on a perfectly clean surface at room temperature for at least 2-3 days in a dust-free atmosphere.
- Put the samples in an oven, and dry at 70°C for 48 hours.
- Grind the samples in an electric stainless steel mill using a 0.5 mm sieve. Clean the cup and blades of the grinding mill before each sample.
- Put the samples back in the oven, and dry again for constant weight. Store in well-stoppered plastic or glass bottles or in paper bags for analysis (Motsara and Roy, 2008).

## Laboratory Techniques for Analysis

As in the case of soil sampling and analysis, plant analysis can serve as a diagnostic tool for farmers. For this, sampling at representative tissues at proper growth stages is very important. Details of plant sampling have been outlined in the preceding paragraphs. This part provides the methodology for analysis of major, secondary and micronutrients in plants.

## Digestion of Plant Samples

The mineral elements from plant tissues are to be released for their determination. Dry-ashing and wet oxidation are the two widely adopted methods. Dry-ashing is a preferred technique particularly for boron and molybdenum analysis. Dry-ashing is carried out in a muffle furnace at an ignition temperature of 550°C followed by its extraction in dilute HCl or $H_2SO_4$. Ashing at temperature above 600°C leads to loss of P, K and B.

Wet oxidation is done using oxidizing acids like $HNO_3$-$H_2SO_4$-$HClO_4$(tri-acid mixture) or $HNO_3$-$HClO_4$(di-acid mixture). Use of $HClO_4$ avoids the volatilization loss of K, provides a clear solution,while $H_2SO_4$ helps to complete oxidation. $HClO_4$ on heating, produces anhydrous $HClO_4$, which dissociates into nascent chlorine and oxygen, which in turn increase the efficiency of oxidation at higher temperature. Di-acid digestion is normally done for the determination of P, K,Ca,Mg,S,Fe,Mn,Zn,Cu. It must be done for the determination of Ca and Mg (Bhargava and Raghupati, 1993). Tri-acid digestion is recommended only when P and K are to be estimated. Sulphur cannot be estimated from tri-acid extract.

Di-acid digestion is carried out using a mixture of 9:4 mixture of $HNO_3$: $HClO_4$. One-half gram or 1.0g dried and processed plant sample is digested in a 100 or 150 ml conical flask. 10mlof Conc.$HNO_3$ is added and a funnel is placed on the flask. This is kept for 6-8 hours or overnight at a covered place for pre-digestion.Then, 10ml of conc.$HNO_3$ and 2-3 ml. of $HClO_4$ is added. This is kept over hot plate in the digestion chamber at 100Ú C for 1 hour, then the temperature is raised to 200°C. Digestion is continued till the contents become colourless and dense fumes appear.The contents are heated for sometime before removing it from hot plate.This is cooled and 10 mlof dilute (approx.2N,AR grade) HCl is added.This is warmed slightly and filtered through Whatman No.42 filter paper into a 100ml volumetric flask.The volume is made up to mark by washing with distilled water.

Tri-acid digestion mixture consists of $HNO_3$,$H_2SO_4$ and $HClO_4$ in the ratio 10:4:1. One-half gram or 1.0 g of plant sample is taken in a 250 ml. conical flask. Five millilitres of conc. $HNO_3$ is added and a glass funnel is placed on the flask. It is placed on a water-bath for about 30 minutes. The flask is then placed over a hot plate at 180-200°C. Boiling is continued to near dryness. Then 5ml. of tri-acid mixture is added. This is heated at 180-200°C, till dense white fumes evolve. Digestion is continued till the mixture is largely volatilized. If the contents are still brown, this is cooled and 3-4 ml. of tri-acid mixture is added and digestion continued. The flask is removed when moist,clear white contents are left. The entire quantity of $HClO_4$ has volatilized at this stage. This is cooled and 50 ml. of distilled water is added.This is then filtered in a 100ml.volumetric flask and the volume is made up to mark.

## Nitrogen Analysis

### *Principle*

Plant samples are placed in a Kjeldahl flask and digested in conc. sulphuric acid with a catalyst mixture at 360-410°C using a digestion assembly. Kjeldahl flask is slowly heated until the initial reaction is over at low temperature. Rate of digestion is increased by using copper sulphate as a catalyst and anhydrous sodium sulphate or potassium sulphate to raise the boiling temperature of $H_2SO_4$. Otherwise,commercially available potassium sulphate-copper sulphate tablets (e.g., Kjeltab) is used. On completion of digestion, the sample is cooled and diluted as concentrated alkali is added to the digest for distillation.The distilled ammonia is quantitatively absorbed in 2%boric acid, containing mixed indicator and titrated against standard acid.

### *Instruments/Apparatus*

- Digestion assembly or digestion chamber having fume exhaust system,if gas burners or hot plates are used.

- Automatic distillation cum titration system or distillation apparatus system.

***Reagents***

- N $H_2SO_4$: Add 2.8 ml of conc.$H_2SO_4$ to about 990 ml of distilled water. Standardise it against standard 0.1*N*NaOH solution.
- Boric Acid (2 or 4%) with mixed indicator: Dissolve 20g of boric acid powder in warm water by stirring and dilute to 1 litre volume.
- Digestion accelerator mixture: Mix 20 parts of anhydrous $Na_2SO_4$ or $K_2SO_4$ with 1 part of $CuSO_4.5H_2O$ or use commercially available tablets.
- Sodium hydroxide 40% solution: Weigh 400g NaOH in 1000 ml Beaker. Then, add 600 ml distilled water under a fume hood. Stir the contents with a glass rod. Cool the contents and transfer in a screw top bottle.

***Procedure***

*A) Digestion on burner*

- Weigh 200 to 500mg of finely ground, dried, 0.5mm sieved plant sample, wrap in a circle of Whatman filter paper and drop it into 800ml Kjeldahl flask.
- Add 20g of digestion accelerator mixture and 35ml of conc.$H_2SO_4$.
- Place the flask on the burner at about 45 degree angle in a fume chamber and heat on low flame for about 30 minutes to avoid too much frothing.
- Raise the flame gradually, keeping it below the level of liquid inside, in order to prevent loss of ammonia and excessive volatilization of $H_2SO_4$. Continue heating until yellow or dark colour totally disappears and does not re-appear on keeping away from the flame.
- Remove the flask and keep aside to cool. After the contents have cooled, add cautiously about 300ml of ammonia free distilled water. Swirl a little to mix and again cool at room temperature or under running tap water.
- Run a blank with identical Whatman paper circle without the plant sample.
- Take 25ml of 4% boric acid containing mixed indicator in a 250ml conical flask and place it under the ammonia receiving tube of the

distillation assembly.

- Add a few glass beads and about 3-4ml of liquid paraffin to the diluted and cooled sample.
- Add 100ml of 40%NaOH solution to the sample and immediately attach to the alkali trap of the distillation unit.
- Continue distillation for about 30-40 minutes collecting 100ml distillate and then remove the conical flask before switching off the heater.
- Titrate against 0.1N$H_2SO_4$ until a purple colour just starts appearing.

*B) Digestion of sample on digestion block*

- Weigh the sample and wrap in filter paper and place in digestion tube. Add one tablet of the digestion accelerator and 6ml of conc. $H_2SO_4$.
- Set the digestion system to attain a temperature of about 385°C and then, attach the digestion tube to the heating unit as per instructions given in operational manual.
- Run the tap water with the desired flow rate (no black or brown colour), which takes about 1(+/- 0.25h).
- Switch off the system.
- Remove the rack of sample tubes along with the exhaust system from the heating unit. Do not stop water flow, as the fumes continue for some more time.
- Set the distillation unit to perform the various steps, *viz*., dilution, addition of alkali, steam generation, titration etc.
- Place boric acid in conical flask and run the distillation for 150 seconds.
- Use the automatic titration system, if available, or do it manually.

***Calculations***

$$\text{Total N in plant (\%)} = \frac{(S-B) \times N \times ME \times 1000}{W}$$

Where,

S is standard acid titre value in ml for the sample distillate.

B is standard acid titre value in ml for the blank distillate.

N is the normality of acid used in titration.

ME is the mg equivalent weight of nitrogen, i.e., 0.014g and

W is the weight of dried plant sample used for digestion.

### *Phosphorus*

*Principle*

Total phosphorus in plant samples is determined in the tri-acid digested material by vanadomolybdo-phosphoric yellow colour method. Vanadate, molybdate and orthophosphates react together to give yellow colour complex in $HNO_3$ medium. Advantage of this method is its extreme simplicity, stability of colour, moderate sensitivity and freedom from interferences. The colour develops in about 30 minutes and is stable for >2 weeks.

*Instruments*

Colorimeter/Spectrophotometer.

### *Reagents*

- Vandomolybdate reagent : Prepare solution 'A'by dissolving 25g ammonium molybdate in about 400ml of warm water. Prepare solution'B' separately by dissolving 1.25g of ammonium metavanadate in about 300ml of boiling water, cool it and add 250ml conc. $HNO_3$. Cool again at room temperature. Now add solution 'A' to solution 'B' and dilute to one litre.
- Phosphorus standard solution: Dissolve 0.2195g of analytical grade $KH_2PO_4$ and dilute to one litre.This solution contains 50μgP/ml (50ppm).

### *Procedure*

- Transfer 5 to 10ml of digested plant material into a 50ml volumetric flask.
- Add 10ml of the vanadomolybdate solution and dilute to 50ml with water.
- Mix well and read the colour intensity after 30 minutes using blue filter (420nm).
- Run a blank without P solution simultaneously.
- Take 0,1,2,3,4 and 5 ml of the 50μgP/ml standard solution to 50 ml volumetric flask to get 0,1,2,3,4 and 5ppm of P, respectively. Add 10

ml vanadomolybdate reagent to each flask. Make up the volume and shake thoroughly. Record absorbance at 420nm. Plot absorbance against concentration.

- Once a linear calibration curve is established,the slope of the curve is determined and then concentration of P in the unknown solution can be calculated as A=mc.
- Absorbance=slope × concentration.

Concentration=Absorbance ÷ slope.

***Calculations***

$$P(\%)=\text{Sample conc.(ppm)} \times \frac{1}{\text{Wt. of sample (g)}} \times \frac{100}{\text{Aliquot (ml)}} \times \frac{\text{Final volume (ml)}}{10000}$$

***Potassium***

Potassium in the acid digested plant material is estimated, with the help of a Flame Photometer either directly or after suitable dilution.

**Standard Solution/Standard Curve**

Dissolve 1.9069g of analytical grade KCl in distilled water and make the volume to one litre. This solution contains 1000ppm K. Prepare 100ppm K solution by diluting the 1000ppm K solution 10 times (10ml in 100ml final volume). Final standard solutions of 0.5 and 10ppm are prepared from 100ppm K solution.

To prepare standard curve, the instrument is set at highest concentration of 5ppm K using a standard filter. The manufacturers specify the linear range of K (normally, 5ppm), and suitable factor is calculated for finding out K in plant samples. The samples are read in flame photometer at 548nm wavelength or using filter of K.

***Calculation***

$$K(\%)=R \times \frac{5}{100} \times \frac{100}{\text{sample weight}} \times \frac{100}{1,000,000}$$

Where, 5ppm K = 100R and if further dilution is made, then appropriate modification is required in the calculation.

***Sulphur***

Since dry ashing leads to volatilization loss of S present in organic combinations, and wet oxidation based on tri-acid mixture includes $H_2SO_4$, both these methods cannot be used for S determination in plant samples. Therefore, $HNO_3$-$HClO_4$ digest of plant samples can be conveniently used. Turbidimetric method as described for soil sulphur determination can be used.

### Ca,Mg,Zn,Cu,Fe and Mn estimation by AAS

Secondary and micronutrients,such as Ca,Mg,Zn,Cu,Fe and Mn are estimated in the di/tri-acid digested plant samples after suitable dilution using Atomic absorption spectrophotometer.

### Boron

Boron estimation in plant extracts is based on the formation of a coloured product (rosocyanin) soluble in alcohol,when solution of $H_3BO_3$, oxalic acid and curcumin are evaporated to dryness at 55± 3°C. At 540nm, absorbance of an alcoholic solution of rosocyanin is proportional to that of B concentration. The procedure is appropriate for B concentrations ranging from 0.1 to 2.0 μgB/ml. Boron can be estimated conveniently by Inductively Coupled plasma (ICP) of recent origin. Soda glass or plastic ware should be used. However, the alcoholic solution of rosocyanin can be handled in borosilicate glassware (Tandon, 2005).

## Interpretation of Leaf Analysis Data

Leaf analysis has been interpreted in several ways. Some of the techniques adopted widely are:

1. Critical value approach
2. Nutrient concentration ranges
3. Diagnosis and recommendation Integrated system (DRIS)
4. Compositional nutrient diagnosis (CND) and Principal component analysis (PCA).

### Critical Value Approach

The concept involves the relationship between the concentration of a single nutrient and growth (yield) of the crop, assuming all other nutrients are adequate. The adequate level of nutrient refers to that point on the growth curve that is about 10% lower than for maximum yield. Similarly, other critical stages: deficiently and toxicity can be worked out. The technique essentially involves finding out the value of X (the critical value), that best divides the data into two populations or classes, from the point of view of prediction using either the graphical (Cate and Nelson, 1965) or the statistical (Cate and Nelson, 1971) method (Fig.2). The range of concentration of the nutrient between the deficient and toxic critical limits can be considered optimum for plant growth. Kotur and Kumar (1989) used this technique to delineate the critical levels of deficiency and toxicity for hot-water soluble B in soil, B concentration of leaf as well as Ca:B ratio in leaf of cauliflower.

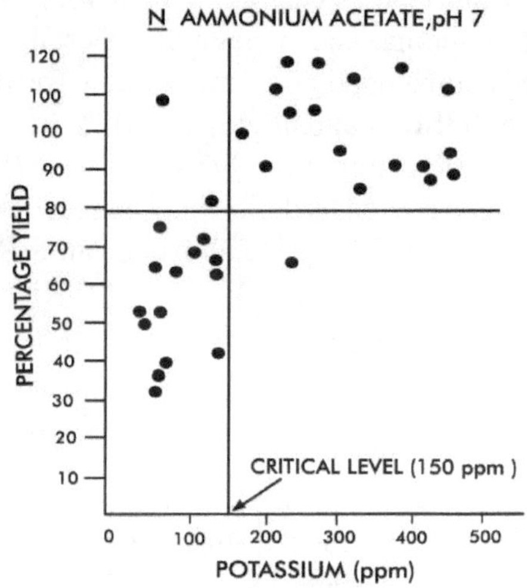


**Fig. 2 :** A typical Cate and Nelson technique applied to identify critical level of available potassium in soil.

The advantage of critical values, once properly established, is their wide applicability for the crop. The disadvantage is that they only provide 'yes or no' type of information and do not cover the entire range, over which nutrient supplies need to be managed. Table 4 presents some critical values for a range of crops.

**Table 4 :** Critical nutrient concentration for 90% yield of crops

| Nutrient | wheat & rice | oilseed rape | sugarcane | alfalfa* | grass | citrus |
|---|---|---|---|---|---|---|
| N (%) | 3.0 | 3.5 | 1.5 | 3.5 | 3.0 | 2.5 |
| P (%) | 0.25 | 0.3 | 0.2 | 0.25 | 0.4 | 0.15 |
| K (%) | 2.5 | 2.5 | 1.5 | 2.0 | 2.5 | 1.0 |
| Mg (%) | 0.15 | 0.2 | 0.12 | 0.25 | 0.2 | 0.2 |
| S (%) | 0.15 | 0.5 | 0.15 | 0.3 | 0.2 | 0.15 |
| Mn (µg/g) | 30 | 30 | 20 | 30 | 60 | 25 |
| Zn (µg/g) | 20 | 20 | 15 | 15 | 50 | 20 |
| Cu (µg/g) | 5 | 5 | 3 | 5 | 8 | 5 |
| B (µg/g) | 6 | 25 | 1.5 | 25 | 6 | 25 |
| Mo (µg/g) | 0.3 | 0.3 | 0.1 | 0.2 | 0.3 | 0.2 |

Ref.Motsara and Roy (2008).Fert. & Plant nutrition Bulletin

*Not critical, but optimal concentration for cows producing 15 litres milk/day

## Nutrient Concentration Ranges

For establishing optimum nutrient concentration ranges, the nutrient concentration *vs.* Yield data is divided into two groups as high and low yielding. The mean value of the nutrient concentration of the high yielding population is called tissue parameter optima. The ranges are then established by considering the mean of the high yielding population as optimum.The range for optimum is then derived by mean-4/3SD (standard deviation) to mean+4/3SD. The range for low is obtained by calculating mean-8/3SD to mean+8/3SD and the value below mean-8/3SD is considered as deficient. The value from mean+4/3SD to mean+8/3SD is taken as high.The value above mean+8/3SD is considered as excess.

The optimum nutrient concentration ranges for important fruit crops under peninsular Indian conditions are presented in Tables 5 and 6.

**Table 5 :** Optimum nutrient norm for fruit crops

| Nutrient | Unit | Anab-e-Shahi | Thompson Seedless | Alphonso | Totapuri | Guava | Acid Lime |
|---|---|---|---|---|---|---|---|
| Nitrogen | % | 0.50-1.25 | 0.87-1.61 | 0.78-1.65 | 0.84-1.53 | 1.69–2.20 | 1.53-2.10 |
| Phosphorous | % | 0.46-0.68 | 0.29-0.65 | 0.02-0.33 | 0.064-0.147 | 0.16–0.23 | 0.10-0.15 |
| Potassium | % | 1.16-2.42 | 2.00-3.02 | 0.77-1.73 | 0.52-1.10 | 1.20–1.67 | 0.96-1.66 |
| Calcium | % | 1.40-2.07 | 0.98-1.36 | 0.76-1.63 | 1.97-3.20 | 0.60–1.27 | 3.05-3.43 |
| Magnesium | % | 0.24-0.58 | 0.63-1.10 | 0.40-0.65 | 0.40-0.65 | 0.35–0.50 | 0.40-0.60 |
| Sulphur | % | 0.08-0.18 | 0.09-0.13 | 0.03-0.13 | 0.014–0.21 | 0.29–0.47 | 0.25-0.29 |
| Iron | ppm | 38-107 | 54-80 | 657-963 | 48 – 86 | 114–178 | 117-194 |
| Manganese | ppm | 18-86 | 42-209 | 13-408 | 57-174 | 34 – 77 | 21-63 |
| Zinc | ppm | 25-94 | 30-88 | 7.8-18.3 | 25-33 | 29 – 41 | 25-50 |
| Copper | ppm | 10-30 | 5-10 | 14.3-17.8 | 3.10-8.00 | 6 – 12 | 8.7-14.8 |

**Table 6 :** Optimum nutrient norms for fruit crops (cont'd)

| Nutrient | Unit | Papaya | Coorg mandarin | Sapota | Pome granate | Robusta | Ney poovan |
|---|---|---|---|---|---|---|---|
| N | % | 0.91-1.44 | 1.82 – 2.72 | 1.60 – 1.85 | 0.91-1.66 | 1.67-3.43 | 2.23-3.35 |
| P | % | 0.12-0.28 | 0.15 – 0.63 | 0.10 – 0.13 | 0.12-0.18 | 0.12-0.21 | 0.12-0.23 |
| K | % | 2.22-3.75 | 1.15 – 2.06 | 1.63 – 1.85 | 0.61-1.59 | 2.28-4.14 | 2.68-4.78 |
| Ca | % | 2.65-4.19 | 0.73 – 2.44 | 0.54 – 0.74 | 0.77-2.00 | 0.48-1.70 | 0.80-1.28 |
| Mg | % | 0.67-1.37 | 0.17 – 0.32 | 0.42 – 0.47 | 0.16-0.42 | 0.33-0.58 | 0.30-0.65 |
| S | % | 0.18-0.37 | - | 0.28 – 0.37 | 0.16-0.26 | 0.03-0.18 | 0.06-0.13 |
| Fe | ppm | 33-92 | 43.19 – 171 | 113 – 161 | 71-214 | 53-196 | 58-189 |
| Mn | ppm | 34-60 | 14.4 – 73.12 | 21 – 31 | 29-89 | 112-417 | 142-516 |
| Zn | ppm | 17-40 | 13.38 – 36.7 | 14 – 17 | 14-72 | 18-38 | 14-37 |
| Cu | ppm | 6-12 | - | 5 –7 | 29-72 | 10 – 32 | 11-33 |

## Diagnosis and Recommendation Integrated System (DRIS)

This holistic approach to the mineral nutrition of crops, integrates the sets of norms representing calibrations of plant tissue, soil composition, environmental parameters and farming practices as the functions of the yield of crop (Beaufils, 1973). Once such derivations are derived, it becomes possible to make diagnosis of the conditions of the crop, thereby isolating the factors, which are likely to be responsible for limiting the growth and production. This approach can make the diagnosis at any stage of the crop development and list the nutrients in order of importance, which are responsible for limiting the yield. DRIS has been developed to fulfil the diagnostic and predictive use of the leaf analysis. Optimizing these factors creates conditions suitable for higher yields and quality.

Development of DRIS norms: In contrast to other approaches, which require experimental data, DRIS employs survey technique using a large number of randomly distributed sites throughout the selected area. At each site, the samples of leaf and soil are collected for analysis and data on applied manures and fertilizers is recorded. The procedure for development of DRIS norms are:

I. Define parameters to be improved and the factors likely to affect them;

II. Collect all the reliable data available from the fields and experimental plots;

III. Study the relationship between the yield and environmental or external factors such as soil composition, weather conditions etc., so that the most favourable condition may be determined in a particular case.

This helps in taking suitable corrective measures on such interrelationship.

IV. Study the relationship between the yield and internal parameters, such as petiole composition using following steps:

a. Each internal plant parameter is expressed in as many forms as possible such as, N/DM, DM/N, N×DM, N/P, P/N, N×P etc.

b. The whole population is divided into a number or subgroups based on the economic yield of the area or zone.

c. The mean of each sub-population is calculated for each the forms of expression.

d. If necessary, class interval limits between the average and the outstanding yields are re-adjusted, so that the means of below average and average yielding populations remain comparable.

e. Chi-square test is performed to know that populations conform to normal distribution.

f. The variance ratios between the yields of sub-populations for all the forms of expression are calculated together with coefficient of variation.

g. The forms of expression, for which significant variance ratios are obtained and the mean values for the population are essentially same, and related to one another in expression with common nutrient.

h. For calculation of DRIS index, Bhargava and Chadha (1993) and Bhargava and Raghupathi (1993) worked out formulae for petiole/leaf N, P, K, Ca, S, Fe, Mn, Cu and yield of grape, mango and acid lime. The formulae developed under one set of condition and for one crop is given below:

$$N = 1/10[f(N/P) - f(K/N) + f(N/Ca) + f(N/Mg) + f(N/S) - f(Fe/N) + f(N/Mn) + f(N/Zn) - f(Cu/N) + f(N/Y)]$$

where, $f(N/P) = [N/P \, / \, n/p - 1][1000/CV]$ when $N/P > n/p$ and

$= [1 - n/p \, /N/P] \, [1000/CV]$ when $N/P < n/p$

And N/P = the actual value of the ratio of N and P in the plant under diagnosis, n/p = the value of the norm (which is the mean value of high yielding orchards) and CV = coefficient of variation for the population of high yielding orchards. Similarly,

$$P = 1/10[-f(N/P) + f(P/K) - f(Ca/P) + f(P/Mg) - f(S/P) - f(Fe/P) + f(P/Mn) + f(P/Zn) - f(Cu/P) + f(P/Y)]$$

$$K = 1/10[f(K/N) - f(P/K) - f(Ca/K) - f(Mg/K) - f(S/K) - f(Fe/K) +$$

$f(K/Mn) + f(K/Zn) - f(Cu/K) + f(K/Y)]$

$Ca = 1/10[-f(N/Ca) + f(Ca/P) + f(Ca/K) + f(Ca/Mg) + f(Ca/S) + f(Ca/Fe) + f(Ca/Mn) + f(Ca/Zn) - f(Cu/Ca) + f(Ca/Y)]$

$Mg = 1/10[-f(N/Mg) + f(P/Mg) + f(Mg/K) - f(Ca/Mg) - f(S/Mg) - f(Fe/Mg) + f(Mg/Mn) + f(Mg/Zn) - f(Cu/Mg) + f(Mg/Y)]$

$S = 1/10[-f(N/S) + f(S/P) + f(S/K) - f(Ca/S) + f(S/Mg) - f(Fe/S) + f(P/Mn) - f(Zn/S) - f(Cu/S) + f(S/Y)]$

$Fe = 1/10[f(Fe/N) + f(Fe/P) - f(Fe/K) + f(Fe/Ca) + f(Fe/Mg) + f(Fe/S) - f(Mn/Fe) - f(Zn/Fe) - f(Cu/Fe) + f(Fe/Y)]$

$Mn = 1/10[-f(N/Mn) - f(P/Mn) - f(P/Mn) - f(K/Mn) - f(Ca/Mn) - f(Mg/Mn) - f(S/Mn) - f(Zn/Mn) - f(Cu/Mn) + f(Mn/Y)]$

$Zn = 1/10[-f(N/Zn) - f(P/Zn) - f(K/Zn) - f(Ca/Zn) - f(Mg/Zn) + f(Zn/S) + f(Zn/Fe) + f(Zn/Mn) - f(Cu/Zn) + f(Zn/Y)]$

$Cu = 1/10[-f(Cu/N) + f(Cu/P) + f(Cu/K) + f(Cu/Ca) + f(Ca/Mg) + f(Cu/S) + f(Cu/Fe) + f(Cu/Mn) + f(Cu/Zn) + f(Cu/Y)]$

$Y = 1/10[-f(N/Y) - f(P/Y) - f(K/Y) - f(Ca/Y) - f(Mg/Y) - f(S/Y) - f(Fe/Y) - f(Zn/Y) - f(Cu/Y)$

One must remember that, the composition of the formulae will depend on the situation and there is no fixed formula to be used under all crop/ site/ situations. The investigator can make an attempt to work out his own for the crop for use.

Diagnosis: Initially standard values or norms were determined from DRIS. These standard values were then used to develop indices by following the equations given above. The value of each ratio function is added to the index sub-total and subtracted from another, prior to averaging. Therefore, all indices are balanced around zero. The higher negative index shows that the core responding nutrient is relatively deficient. Alternatively, a large positive index indicates the nutrient is excessive in quantity. An example of identification of yield limiting nutrients through DRIS indices in fruit crops is illustrated below (Table 7 and 8)

**Table 7 :** Diagnosis of Nutrient imbalance using DRIS

| Parameter | Limiting | | | | Optimum | | | Excess | | Imbalance | |
|---|---|---|---|---|---|---|---|---|---|---|---|
| | N | Mg | Zn | Ca | K | S | P | Cu | Mn | Fe | Sum |
| Nutrient Concentration* | 1.07 | 0.26 | 30 | 1.26 | 1.34 | 0.12 | 0.52 | 14 | 140 | 118 | 1422 |
| DRIS Indices | -269 | -197 | -149 | -96 | 5 | 64 | 67 | 147 | 158 | 270 | |

*Concentration of N, P, K, S, Ca and Mg expressed in (%).

Concentration of Fe, Cu, Mn and Zn expressed in ppm.

**Table 8 :** DRIS norms derived from high yielding population for Banana

| Nutrient ratio | Mean | CV | Nutrient ratio | Mean | CV |
|---|---|---|---|---|---|
| P/K | 0.064 | 30.22 | K/Fe | 0.0348 | 42.96 |
| K/N | 1.364 | 25.04 | Mn/K | 93.52 | 53.18 |
| Ca/N | 0.303 | 38.01 | Zn/K | 7.283 | 39.60 |
| N/Mg | 9.639 | 71.40 | Ca/Mg | 2.720 | 71.86 |
| S/N | 0.035 | 31.12 | Ca/S | 9.305 | 51.25 |
| N/Fe | 0.026 | 41.12 | Ca/Fe | 0.0078 | 54.82 |
| Mn/N | 121.8 | 47.08 | Ca/Mn | 0.0029 | 56.91 |
| Zn/N | 9.57 | 35.88 | Ca/Zn | 0.0036 | 57.00 |
| P/K | 0.050 | 32.74 | S/Mg | 0.3530 | 95.84 |
| Ca/P | 4.986 | 46.71 | Fe/Mg | 400.4 | 64.54 |
| P/Mg | 0.716 | 65.37 | Mn/Mg | 1223.70 | 81.05 |
| P/S | 1.972 | 37.00 | Zn/Mg | 95.69 | 77.98 |
| P/Fe | 0.001 | 57.56 | S/Fe | 0.00093 | 50.18 |
| Mn/P | 1996.9 | 60.00 | Mn/S | 3654.5 | 53.53 |
| Zn/P | 156.7 | 48.06 | Zn/S | 294.16 | 52.37 |
| Ca/K | 0.242 | 57.03 | Mn/Fe | 2.952 | 45.77 |
| K/Mg | 13.57 | 81.06 | Zn/Fe | 0.236 | 45.15 |
| S/K | 0.027 | 39.12 | Mn/Zn | 13.38 | 41.58 |

## Compositional Nutrient Diagnosis (CND)

Nutrient ratios in plant are diagnosed using nutrient concentration or dual ratio in selected tissue (Walworth and Sumner 1987) as explained above. The use of dual ratio in DRIS instead of nutrient concentration could reduce effect of nutrient interaction, dilution or accumulation in plant tissues. However, when multi-nutrient interactions exist in plant, it cannot be explained by DRIS effectively. The compositional nutrient diagnosis (CND), which has been proposed for nutritional diagnosis includes physiological concept, such as nutrient interaction in plant and also mathematical constraint to a closed system known as bounded sum constraint (Parent and Dafir, 1992).

The CND norms were developed by adopting the procedure as outlined by Parent *et. al.* (1994). The full composition array for the nutrient proportions (D) in plant tissues was described by the following simplex ($S^D$) contained to 100%:

$$S^D = [(N, P, K, \ldots R): N>O, P>O, K>O, \ldots, R>O; \; N+P+K+\ldots+R = 100\%]$$

where, 100% is the dry matter content (i.e., the invariable sum of all the components or the full relative composition of the diagnostic tissues). N, P, K are the nutrient concentrations and R is the filling value between 100% and sum of the nutrient concentrations. The value of R is thus, composed of undetermined components as well as experimental error, and was required to linearize compositional data.

The bounded sum constraint to 100% of compositional data was alleviated by correcting nutrient concentrations by geometric mean (G) of all the D components including R.

$$G = [N \times P \times K \times \ldots\ldots \times R]^{1/D}$$

Row centred log-ratios were generated for $V_N$ to $V_{Zn}$ as follows :

$$V_N = \ln(N/G), \ldots\ldots, V_{Zn} = \ln(Zn/G)$$

Expressions such as N/G,... Zn/G are multi-nutrient ratios, since each nutrient is divided by geometric means of all the components (the determined nutrients and the filling value). The row-centered log-ratios are linearized (undistorted) estimates of the original components, that are fully compatible with PCA.

$V^*_N$ to $V^*_{Zn}$ and $SD^*_N$ to $SD^*_{Zn}$ are the CND norms (indicated by asterisks), i.e., mean and standard deviation of each row-centered log-ratios in the high yielding population. The standardized variables $(V_N - V_N^*) / SD^*_N$ to $(V_{Zn} - V^*_{Zn}) / SD^*_{Zn}$ are the CND nutrient indices.

Indices

$$I_N = (V_N - V^*_N) / SD^*_N, \ldots\ldots I_{Zn} = (V_{Zn} - V_{Zn}^*) / SD^*_{Zn}$$

Independent values for $V_N$ to $V_{Zn}$ were introduced in the equation for diagnostic purpose.

**Application of multivariate technique for diagnosis of nutrient imbalance in fruit crops**

The application of CND involves establishment of data bank similar to that of DRIS technique. Recently, the technique has been made use of for establishment of nutrient norm and for identification of yield limiting nutrients in fruit crops like pomegranate and grape (Raghupathi and Bhargava 1998, Bhargava and Raghupathi 1999). The mean and range of nutrient concentration of such a data developed under peninsular Indian condition is shown in Table 9.

The observation of the data indicates, that the concentration of different nutrients is not varying largely in high and low yielding population during flowering stage in case of grape. Only, the concentration of K and Ca is marginally higher in the low yielding population. The concentration of different nutrients during bud differentiation stage also differed only marginally between high and low yielding population (Bhargava and Sumner, 1987). In fact, the mean concentration of N and K is sometimes marginally higher of in low yielding population. This indicates that to draw conclusion merely based on the absolute nutrient concentration is often misleading.

The V values presented are nothing, but the multivariate norms for nutrients. It can be seen that except for K, the norm value is higher during bud differentiation stage compared to flowering stage indicating greater requirement of nutrient during bud initiation and differentiation. These norm values can be reconstituted as shown in Table 9. The reconstituted values are approximately 1.19 and 0.97% N at BUD and FS respectively, indicating that the optimum requirement of N is higher during BDS and that there is a consequent remobilization of N from vines to inflorescence. Large differences in CND norm are noticed for Ca and Mg in grape. The reconstituted Ca and Mg norm values are 2.46 and 0.91%, 0.57 and 0.25% for BDS and FS, respectively. Anab-e-Shahi is a seeded cultivar of grape. And bulk of Ca and Mg is mobilized from vegetative organ for flowering and subsequent development of seeds in fruits. The Zn concentration also varies widely in grape during bud differentiation and flowering.

**Table 9 :** Compositional nutrient diagnosis norms (mean and deviation) of row centred log-ratios for grape

| Row centered log ratio | Bud differentiation stage | | | Flowering stage | | |
|---|---|---|---|---|---|---|
| | **Mean** | **Concentration*** | **SD** | **Mean** | **Concentration*** | **SD** |
| $V_N$ | 1.65 | 1.19 | 0.320 | 1.38 | 0.97 | 0.270 |
| $V_P$ | 1.26 | 0.81 | 0.170 | 0.90 | 0.56 | 0.375 |
| $V_K$ | 2.38 | 2.48 | 0.218 | 2.72 | 3.49 | 0.301 |
| $V_{Ca}$ | 2.37 | 2.46 | 0.135 | 1.38 | 0.91 | 0.433 |
| $V_{Mg}$ | 0.91 | 0.57 | 0.228 | 0.11 | 0.25 | 0.328 |
| $V_S$ | 0.21 | 0.18 | 0.239 | 0.48 | 0.14 | 0.480 |
| $V_{Fe}$ | 3.16 | 97 | 0.395 | 3.61 | 62 | 0.746 |
| $V_{Mn}$ | 3.65 | 59 | 0.460 | 4.05 | 40 | 0.567 |
| $V_{Zn}$ | 3.38 | 78 | 0.460 | 4.23 | 33 | 0.537 |

* Reconstituted nutrient concentration

The values for N, P, K, Ca, Mg, S are in % and Fe, Mn, Zn in ppm.

**Principal Component Analysis (PCA)**

A reasoned application of principal component analysis could lead to the greater understanding of the effect of fertilization treatments on leaf composition. The principal component analysis reduces the number of interdependent variables into smaller number of independent principal components, that are linear combinations of original variates. Principal component analysis was performed on log transformed nutrient concentration data prior to statistical computation that followed normal distribution. To be declared as significant, principal components must have Eigen values >100/P, where P is the total number of varieties under diagnosis. Alternatively, principal components showing Eigen values <1 were considered non-significant. Only, principal component loading in Eigen vectors having values greater than the selection criterion (SC) are given significance. The selection criterion was computed as follows:

$$SC = 0.50 / (\text{PC Eigen values})^{0.5}$$

The principal component analysis conducted on log-transformed data produced four significant principal components explaining about 70.30 % of variance (Table 10) under protected conditions. Since, principal components are linear contrasts among the nutrients, interpretation of principal components should consider the sign of the variate. The first principal component was positively correlated with N and P and negatively correlated with Fe and Zn and designated as ($N^+P^+Fe^-Zn^-$) indicating that N and P behaved in one direction, Fe and Zn in the opposite direction under protected conditions. In the second principal component, the antagonistic effect between Ca and S was evident and was designated as ($Ca^-S^+$). In the third principal component, K was positively correlated, while Fe and Mn behaved in the opposite direction. In the fourth principal component, P and Mg behaved in opposite direction, indicating that build up of Mg resulted in substantial decrease in P concentration under protected conditions. The loading for N was significant only in the first principal component indicating that N was associated with P, Fe and Zn, and had least interaction with other nutrients.

Under open conditions, the principal component analysis conducted on log transformed nutrient concentration data showed only two significant principal components explaining nearly 74.26% of the variance (Table 10). The first principal component ($N^+P^+Ca^-Mg^-S^+Fe^-Mn^-Zn^+$) explained nearly 53.27 % of the variance, although the full nutrient interaction was not explained by it alone but as many as five nutrients ($K^+Ca^+Mg^+Fe^-Zn^-$) showed significant interaction in principal component. Excluding K, all other nutrients showed a significant loading in principal component. Nitrogen, P, S and Zn formed one group, while

Ca, Mg, Fe and Mn formed the other group. Although it is difficult to interpret such nutrient interactions on physiological basis, the influence of K appears to be much less critical in open conditions, where as K had overwhelming influence under protected conditions with other nutrients. The compositional nutrient diagnosis multivariate technique revealed that, application of major nutrients could result in greater imbalance of Fe and Zn under protected conditions, where as it resulted in greater imbalance of Ca, Mg, Fe, Mn and Zn under open conditions. Thus, the multi-nutrient interactions need to be understood for proper management of nutrients.

**Table 10 :** Principal component analysis loadings performed on log-transformed data

| | **Protected conditions** | | | **Open conditions** | | |
|---|---|---|---|---|---|---|
| Variate | PC1 | PC2 | PC3 | PC4 | PC1 | PC2 |
| N | 0.805* | 0.265 | 0.065 | 0.138 | 0.907* | -0.110 |
| P | 0.489* | 0.346 | 0.000 | -0.606* | 0.882* | -0.209 |
| K | -0.093 | 0.171 | 0.784* | -0.191 | -0.075 | 0.854* |
| Ca | 0.131 | -0.943* | 0.216 | -0.060 | -0.530* | 0.662* |
| Mg | 0.059 | -0.009 | -0.018 | 0.932* | -0.450* | 0.847* |
| S | 0.172 | 0.662* | 0.252 | -0.277 | 0.835* | -0.223 |
| Fe | -0.537* | 0.138 | -0.508* | -0.018 | -0.633* | -0.633* |
| Mn | -0.069 | 0.105 | -0.750* | -0.131 | -0.573* | 0.077 |
| Zn | -0.628* | 0.233 | 0.098 | 0.244 | 0.749* | -0.457* |
| Eigen values | 2.166 | 1.689 | 1.27 | 1.145 | 4.795 | 1.889 |
| % variance | 24.70 | 18.76 | 14.12 | 12.72 | 53.27 | 20.98 |
| Selection Criteria | 0.339 | 0.384 | 0.443 | 0.467 | 0.228 | 0.363 |

## References

Beaufils, E.R. 1973. Diagnosis and recommendation integrated system (DRIS), *Soil. Sci. Bull. No. 1,* University of Natal, South Africa.

Bhargava, B.S. and Chadha, K.L.1993. Leaf nutrient guide for fruit crops. pp. 973-1029 In: Advances in Horticulture Vol. 2 (Chadha, K.L. and Pareeek, O.P. Eds.), Malhotra Publishing House, New Delhi.

Bhargava, B.S. and Raghupathi, 1993. Analysis of plant materials for macro and micronutrients. pp. 49-82. In H.L.S. Tandon (Ed.) *Methods of analysis of soils, plants, water and fertilizers.* Fertilizer Development and Consultation Organization, New Delhi.

Cate R. B and Nelson L A 1971. A simple statistical procedure for partitioning soil test correlation data in to two classes *Soil Sci. Soc. Amer Proc.* 35: 658 – 659.

Cate, R.B. and Nelson, L.A. 1965. A rapid method for correlation of soil test analyses with plant response data. *Technical Bulletin No. 1, ISTP Series,* North Carolina Agricultural Experiment Station, NC.

Jones *et al.* 1991. Plant analysis Handbook. Micro-Macro Publishing, Athens, GA, USA.

Kotur, S.C and Kumar, S. 1989. Response of cauliflower (*Brassica oleracea* convar *botrytis* var *botrytis*) to boron in Chotanagpur region, *Indian Agricutlural Sci.* 59: 640-644.

Motsara, M.R. and Roy, R.N. 2008. Guide to laboratory establishment for plant nutrient analysis. FAOFert. & Pl.Nutri. Bull.19, FAO, Rome, pp1-204.

Parent, L.E. and Dafir, M. 1992. A Theoretical concept of compositional nutrient diagnosis. *Am. Soc. Horti. Sci.* 117: 239-242.

Parent, L.E., Camboruris, A.N. and Muhawenimana, A. 1994. *Soil Sci. Soc. Am. J.* 58: 1432.

Parent, L.E.; Isfan, D.;Tremblay, N.; Karam, A. 1994. Multivariate Nutrient Diagnosis of carrot crop. *J. Am. Soc. Hort. Sci.*, 119: 420-426.

Raghupathi, H.B.; Bhargava, B.S. 1998. Diagnosis of Nutrient Imbalance in Pomegranate by Diagnosis and Recommendation Integrated System and Compositional Nutrient Diagnosis. Commun. *Soil Sci. & Plant Anal.*, 29: 2881-2892.

Reuter, D.J. and Robinson, J.B. 1986. Plant analysis-an interpretation manual. Inkatapress. melbourne, Sydney, Australia.

Tandon, H.L.S. ed. 2005. Methods of Soils, Plants, Waters, Fertilisers and Organic Manures. F.D.C.O., New Delhi. pp1-204.

❑❑❑

CHAPTER - 6

# Water Analysis

**G. N. Chattopadhyay and Abira Banerjee**

In agricultural systems, water quality is usually determined to assess its suitability for irrigation and/or its use for aquaculture. Analysis of irrigation water is carried out, to know their nutrient load, suitability for irrigation purposes in general and to salt sensitive crops in particular, and their significance as nutrient suppliers as in practical crop production (Tandon,1993). For aquaculture, on the other hand, maintenance of pond environment for survival and growth of aquatic lives, forms the major purpose of such exercises. While both the assessments are mostly governed by some common variables, a few additional attributes are also used for identifying the water quality for these two different purposes. The most important characteristics, that determine the quality of irrigation water are:

- pH
- total concentration of soluble salts assessed through EC
- relative proportion of Na to Ca and Mg, referred to as SAR.
- concentration of B and other elements toxic to plants
- concentration of $CO_3$ and $HCO_3$ as related to Ca+Mg, referred to as RSC
- content of anions such as Cl, $SO_4$ and $NO_3$

On the other hand, for assessing the water quality in aquaculture systems, some additional properties are generally given more emphasis. They are:

- dissolved oxygen
- free carbon-di-oxide
- alkalinity
- biological oxygen demand.

In both the cases, evaluation of water quality is carried out under a few common stages viz. collection of water samples, preservation, analysis and interpretation.

## 1. Collection of Water Samples

Collection of representative water sample is an integral part of the total analytical procedure and it should be given more importance than is usually done. Generally, a small amount of water sample is collected from the source of water for various analyses. Every effort should be made so that this sample becomes a true representative of the water resource.

### *1.1 Procedures of water sampling*

Water samples are commonly collected in stoppered plastic or glass bottles of different sizes depending on the requirement of samples. The bottles should be thoroughly cleaned to avoid any contamination. The bottle should be rinsed a few times with the water to be sampled and then the water is to be filled in the bottle. In case of fish pond water, some attention is to be paid on the location of sampling also. Maximum photosynthetic activity of primary fish feed organisms are observed in the upper layers of pond water and hence this zone should be given major attention. Hour of sampling is another important criterion to be considered for pond water. Due to the effect of sunlight on photosynthetic activity in a fish pond, dissolved oxygen (DO) levels as well as pH values of water usually tend to increase during day time whereas, the amount of free carbon di-oxide ($CO_2$) decreases. Hence, time for sample collection should be kept fixed for all regular sampling activities, because deviation from routine hour of collection may not yield comparable results.

For many specific purposes, it often becomes necessary to know the water quality at some particular depths. This can be achieved by using different water samplers generally available in the market. One such commonly used water sampler is "Kemmerer's Sampler", various modifications of which have been and may be made for different purposes. This type of water sampler consists of à container, the lid of which can be closed with the help of a messenger after filling in water from the desired depth. As the sampler is pulled up, the collected water sample is transferred to the collection bottle through a draining pipe fitted with a clamp. After collection, the bottle is pulled up along with the sampler, removed from the structure and the collected water sample is used for analysis.

### *1.2 Preservation of water samples*

Some of the water properties change quite rapidly on storage and hence adequate care should be taken to complete the water analysis as early as possible after collection. Boyd (1978) described that number of changes may occur among the constituents of water in sampling bottles including the absorption of substances in the sides of the bottles, metabolic activities of micro organisms and the release of metabolites by them. This may lead to alteration of pH, changes in DO and $CO_2$ concentrations, variations in alkalinity, transformation of different forms of nitrogen (N), viz. ammonification, nitrification, denitrification etc. and some other properties of water. However, it is not always possible to complete the analytical processes within a short period, especially when water samples are collected from places away from laboratory. Different preservation methods are available for inhibiting changes in water qualities for a considerable period, which provides some more time to the analyst for completing his analyses. Some of guidelines for using preservatives for different analytical procedures havå been presented in Table 1.

**Table 1 :** Use of preservatives in chemical analyses of water

| Parameter | Preservative | Permissible time |
|---|---|---|
| Acidity and Alkalinity | Cooling at 4°C | 24 hours |
| Biochemical Oxygen Demand | do | 6 hours |
| Carbon Di-oxide | do | 2 hours |
| Chlorophyll-a | do | 12 hours |
| Chemical Oxygen Demand | 1 ml conc. $H_2SO_4$ $l^{-1}$ | 7 days |
| Hardness | 1 ml conc. $HNO_3$ $l^{-1}$ | 7 days |
| Ammonium and Nitrate | 1 ml conc. $H_2SO_4$ $l^{-1}$ | 7 cays |
| Soluble Phosphorus | Storage in $I_2$ treated bottles | 7 days |
| Dissolved oxygen (Winkler's method) | Fixing With Manganous Sulphate and Alkaline Iodide immediately | 6 hours |

(Chattopadhyay, 1998)

## 2. Analysis of water sample

In this section, analytical procedures of some commonly used water quality attributes have been discussed. Number of methods are available for estimation of most of these properties. However, considering that most of the laboratories engaged in water analysis in this country are not usually very well equipped, attempts have been made to discuss the more common as well as simple methods for these determinations.

### *2.1. Temperature*

Water temperature is determined mostly in fish ponds and also for different effluents. The general procedure for determination of water from open water bodies has been discussed here.

*Principle*

For determining the water temperature in open water, a centigrade thermometer graduated at $0.1^0$ C interval is usually used. If more accuracy is sought for, a thermometer with $0.01^0$C divisions may be selected. For subsurface water, a specially designed reversing thermometer is used, where the temperature reading can be fixed after the thermometer comes in equilibrium with water temperature.

*Requisites*

Thermometer : Centigrade thermometer with divisions as per requirement

*Procedure*

To determine the temperature of water, dip the thermometer directly into the water, keep it there for about one minute and note the temperature. Repeat the procedure a few times. In case of subsurface water samples, use the reversing thermometer. Place the reversing thermometer at the desired depth under water and, reverse the thermometer with the help of a trigger arrangement which will help the mercury level to get fixed. The thermometer is taken out and the temperature reading is noted.

### *2.2. pH*

pH is probably the most attended attribute in water quality analysis. While pH is primarily estimated for assessing the acidity or alkalinity of water, its impact on various chemical as well as biological properties of water are also well known. This property of water may be measured either colorimetrically or potentiometrically. The colorimetric methods involve less expenditure but may also be less accurate and need more time for determination than the electrical method (Golterman *et.al.* 1978). However, with the development of battery operated potentiometers, which can be used at field levels, colorimetric determination of pH has now become almost obsolete.In this communication, therefore, only the potentiometric pH measurement of water has been discussed.

*Principle*

In potentiometric determination of pH, electromotive force of a galvanic cell is related to activity of hydrogen ion in water. The voltage produced by the cell is converted by the potentiometer to a current, which is proportional to the pH of the sample. The current is transmitted in terms of pH reading.

*Requisites*

(a) Apparatus – pH meter

(b) Glasswares – 100 ml beakers, 100 ml volumetric flasks

(c) Reagents – Buffer solutions

Standard buffer tablets for different pH levels are available in the market. Three buffer solutions having pH 4.0, 7.0 and 9.2 are commonly used for calibrating pH meter. Usually, one buffer tablet dissolved in 100 ml distilled water (DW) produces corresponding pH level of the solution.

*Procedure*

The operational procedure of the pH meter should follow the guidelines of the manufacturer. Some basic features of such operations have been presented here.

Keep the pH meter in "switch on" position for some time before use to allow a warming up time for the instrument. Then, take about 50 ml of pH 7.0 buffer solution in a clean 100 ml beaker, dip the electrodes (electrode, if combined glass and calomel electrode is used) and calibrate the pH reading to 7.0. Remove the beaker, rinse the electrode (s) with DW, soak any adhering water in a filter paper and dip in to another buffer solution or either acid or alkaline range, depending on the expected pH value of the water sample to be measured. Check if the second solution is giving the respective reading of the buffer. If any fluctuation is observed, calibrate the pH reading with the second buffer and again observe the pH reading in first buffer solution by following the same method. In case water samples of wide range of pH values are to be analysed, the instrument may be calibrated with both ends of pH values (4.0 and 9.2).

When all the buffer solutions are giving exact reading of respective pH values, dip the electrode(s) in about 50 ml of water sample in clean beaker and find the pH reading directly. In case of more than one sample, rinse the electrode(s) every time with DW, soak the moisture with a filter paper and go for next analysis. If large numbers of water samples are to be analysed, some intermediate checking of the calibration is necessary. However, now a days, most of the instruments have a monitoring switch, which indicates whether the calibration has changed or not. After completion of the work, wash the electrode (s) with DW and keep dipped in a beaker containing DW.

*Comments*

A common problem generally encountered in pH estimation is development of sluggishness of the electrode after prolonged use. In such case, the electrode should be dipped in 0.1 N hydrochloric acid (HCl) overnight, thoroughly washed and then used for further work. For battery operated pH meters, it should be ensured that power supplying capacity of the battery is intact.

## *2.3 Electrical Conductivity (EC)*

Electrical conductivity (EC) value indicates that, total concentration of ionized constituents of a water sample and is used as an index of salt content of water. EC is reciprocal of electrical resistance and has been described as specific conductivity of a solution as the conductance that would be measured at $25^0$ C between electrodes 1 $cm^2$ in cross section and placed 1 cm apart and may be visualized as the conductance across a $cm^3$ or mmhos $cm^{-1}$ (Jackson, 1971). Recently, the unit of conductance has been modified to $dSm^{-1}$.

*Principle*

EC is determined with the help of an EC bridge, also known as conductivity meter or EC/TDS analyser. This instrument has a conductivity cell. The cell constant of this cell is determined first by measuring the conductance of standard potassium chloride (KCl) solution, specific conductance of which solution is known.

Cell constant = Known sp. conductance of standard KCl solution/ Conductance shown by the cell.

The conductance of the water sample is measured with this standardized instrument and the obtained conductance value multiplied by cell constant, gives the EC of the sample.

*Requisites*

(a) Apparatus: Conductivity bridge.

(b) Glass Wares: Vol. Flasks 1litre, beakers 100 ml etc.

(c) Reagents:

Standard (0.02M) Potassium Chloride: Dissolve 1.4912 g KCl in 1litre of DW. EC of this solution is 2.39 $dSm^{-1}$ at $18^0$ C and 2.768 $dSm^{-1}$ at $25^0$ C.

*Procedure*

(a) Immerse the cell of the conductivity meter in the standard KCl solution and read the conductivity value in the meter.

(b) Calculate the constant from the relationship shown earlier using known specific conductance of the solution.

(c) Rinse the cell with DW and then immerse in the test water sample, read the conductivity as before.

*Calculations*

Calculate the EC of the water sample from the relationship EC ($dSm^{-1}$) = measured conductivity ($dSm^{-1}$) × Cell constant.

*Comments*

EC values are closely related with temperature. EC of a solution increases approximately at the rate of 2 per cent per $^0C$. Hence, EC is generally expressed at a standard temperature of $25^0C$. Now a days, most of the EC meters are having temperature compensation arrangement, which provides direct EC reading for $25^0C$ .

## *2.4 Total solids*

Total solids of a water sample represent dissolved organic matter, particulate organic matter, dissolved inorganic substances excepting gases and suspended inorganic substances. Different forms of solids present in water eg. total solids (TS), total dissolved solids (TDS) total suspended solids (TSS), total volatile solids (TVS) and total volatile dissolved solids (TVDS) can be measured by following the major principles of evaporation of the water sample.

*Principle*

TS of a water sample can be determined simply by evaporating a known amount of water sample and weighing the residue while TDS can be measured by the same manner but by using filtered water sample only. On the other hand, TSS is the difference between TS and TDS. TVS can be determined from the weight loss after ignition of the residue of TS analysis, while TVDS can be determined by igniting the residue for TDS estimation.

For getting very accurate results for TDS or TVDs, glass fibre filters are recommended. However, in absence of such facility, use of Whatman no. 42 filter paper will yield fairly good results. The second provision has been described in this chapter.

*Requisites*

(i) General: Porcelin basin, Whatman no 42 filter paper, funnel, conical flask etc.

(ii) Muffle furnace.

*Procedure*

For determination of TS, heat a clean porcelin basin in an oven at 105°C. Cool in a desiccator and weigh it accurately ( $W_1$). Shake the water sample thoroughly and take 100 ml of the sample in the basin. Evaporate the water sample to dryness in oven at temperature around $100^0$ C. After drying, keep thå basin at $105^0$ C for half an hour. Cool it in dessicator and weigh the basin along with the residue ($W_2$).

For estimation of TDS, filter the water sample with the help of a Whatman no. 2 filter paper and use 100 ml of the filtrate for drying and weighing ($W_3$ ) as before. TVS can be determined by igniting the residue for TS estimation in a muffle furnace at $550^0$C for half an hour. Cool the basin in a dessicator and weigh it accurately ($W_4$).

Similarly for estimation of TVDS, ignite the residue in the basin for TDS analysis as above and get the weight of the ignited residue ($W_5$).

*Calculation*

TS (mg $kg^{-1}$) = ($W_2$ - $W_1$) mg x 10

TDS (mg $kg^{-1}$) = ($W_3$ - $W_1$) mg x 10

TVS (mg $kg^{-1}$) = ($W_2$ – $W_4$) mg x 10

TVDS (mg $kg^{-1}$) = ($W_3$ – $W_5$) mg x 10

### 2.5. Alkalinity

The amount of acid required to titrate the bases in water is a measure of alkalinity (Boyd, 1978). Number of bases like hydroxides, carbonates, bicarbonates, ammonia, silicate, phosphate etc. may contribute to alkalinity of water. However, most of these bases usually occur in very small concentrations in water bodies to exhibit considerable influence on water qualities. Hence, for practical purposes, presence of hydroxide ($OH^{-1}$), carbonate ($CO_3^{=}$) and bicarbonate ($HCO^-$) are usually considered for determination of alkalinity.

***Principle***

Water samples containing measurable $OH^{-1}$ and / or $CO_3^{=3}$ ions turn pink to phenolphthalein indicator. When titrated with acid to convert these ions to $HCO_3^-$ form, this water becomes colourless at pH below 8.4. Again, water sample with only $HCO_3^-$, can be titrated to the critical pH level of 5.3 with acid by achieving yellow to faint orange colour change of methyl orange indicator. Both the indicators are used for determining total alkalinity, expressed as mg $l^{-1}$ of calcium carbonate ($CaCO_3$) and also the concentrations of $OH^-$, $CO_3^{=}$ and $HCO^-_3$ ions in water samples.

### *Requisites*

(a) Glass wares etc: 1 litre volumetric flasks, 250 ml conical flasks, burette, 100 ml cylinder etc.

(b) Reagents

(i) Phenophthalein indicator. Dissolve 0.5g phenolphthlein in 50 ml of 95 per cent ethyl alcohol and add to it 50 ml DW.

(ii) Methyl orange indicator (MO). Dissolve 0.5g methyl orange in 100 ml DW.

(iii) Standard (0.02 N) Sulphuric acid ($H_2SO_4$). Prepare 1 litre of 0.02 N $H_2SO_4$ by taking 0.55 ml of conc. $H_2SO_4$ and standardized with a primary standard viz., $Na_2CO_3$.

*Procedure*

Take 100 ml water sample in a 250ml conical flask and add 3 drops of phenolphathlein indicator. If the water turns pink, titrate with 0.02 N $H_2SO_4$, until the pink colour just disappears. Note the volume (ml) of acid used. Then add 3 drops of MO indicator in the same (used) water sample. If the water turns yellow, titrate with the same acid, until a faint orange end point is obtained. Note the volume (ml) of the acid used during this titration.

*Calculation*

If the volume (m) of 0.02 N $H_2SO_4$ used for titration with phenolphthalein is P and the total volume of 0.02 N $H_2SO_4$ consumed during titration with both phenolphthalein and methyl orange is M, total alkalinity as ppm of $CaCO_3$, will be M x 10.

Theroux *et al* (1943) described following conditions for the purpose of estimation of concentrations (mg $l^{-1}$) of $OH^-$, $CO^=_3$ and $HCO^-_3$, ions in water in terms of $CaCO_3$ equivalent.

(i) if P= M, $OH^- = P \times 10$

(ii) if P> ½ M, $OH^- = (2P - M) \times 10$

$CO^=_3 = 2(M - P) \times 10$

(iii) if P = ½ M, $CO^=_3 = M \times 10$

(iv) if P< ½ M, $CO^=_3 = 2P \times 10$

$HCO^-_3 = (M-2P) \times 10$

(v) if P = 0 $HCO^-_3 = M \times 10$

***Comments***

Forms of alkalinity may change appreciably on storage. Hence, water samples should be analysed soon after collection, if forms of àlkalinity are of major interest. Samples for total alkalinity may, however, be stored for a few days without appreciable changes.

## *2.6. Chloride*

Most of the water soluble salts in water generally remains in $Cl^-$ form and hence the amount of $Cl^-$ in a water indicates very closely the total amount of soluble salts present .

*Principle*

$Cl^-$ ions are commonly estimated by titration with silver nitrate ($AgNO_3$) in presence of chromate ions. $AgNO_3$ forms silver chloride (AgCl) by reacting with the $Cl^-$ ions present in water. When the Cl in water becomes exhausted, $AgNO_3$ reacts with the $CrO^{=}_4$ ions to show a red colour of silver chromate ($AgCrO_4$) indicating that, the titration has been completed.

*Requisites*

(a) Glass wares, Burette Pipette, volumetric flasks, Porcelin basin etc

(b) Reagents

i) Potassium Chromate ($K_2CrO_4$) indicator: Dissolve 5g $K_2CrO_4$ in about 75 ml of water and add dropwise a saturated solution of $AgNO_3$ to form a red precipitate completely. Filter the suspension ànd dilute the filtrate to 100 ml.

ii) Standard Silver Nitrate ($AgNO_3$) solution: Weigh exactly 4.791 g $AgNO_3$ and dissolve in 1 litre of DW. One ml of this solution will be equivalent to 1 mg of Cl. Standardize the strength of the solution by titrating against a standard sodium chloride (NaCl) solution, by following the procedure part of this section.

iii) Standard Sodium Chloride (NaCI) solution: Dissolve exactly weighed 1.618 g dry, analytical grade NaCl in 1 litre of DW. 1 ml of this solution will contain 1 mg of Cl.

***Procedure***

(a) Standardization of $AgNO_3$ solution

Take10 ml of standard NaCl solution in a porcelin basin. Add DW to make the volume around 100ml, Add 0.5ml of $K_2CrO_4$ to the solution and titrate with $AgNO_3$, till appearance of a permanent red precipitation. Standardise the strength of $AgNO_3$ solution with the help of titre value.

If F be the factor of the $AgNO_3$ concentration and V the volume required for titration then F=10/V.

(b) $Cl^-$ estimation

Take 20-100 ml of water sample in a porcelin basin, add DW if necessary, to increase the volume ànd proceed as above to get the end point at formation of permanent red precipitation.

*Calculation*

If X be the titre value (ml) of $AgNO_3$ , F the factor and V the volume of water sample (ml)

$$\text{Concentration of } Cl^-(\text{mg } l^{-1}) = \frac{X \times F \times 1000}{V}$$

*Comments*

During titration of water sample with $AgNO_3$, the permanent red coloured precipitation can be visualised only after the reaction between $Cl^{-1}$ and $AgNO_3$ has been completed. Hence, for getting more accurate titre value, Theroux et al. (1943) suggested that a value of 0.2 may be deducted from the observed burette reading and then used for calculation.

## *2.7. Sulphate*

Sulphate ($SO_4$) is not commonly included on regular ànalysis of water, either for irrigation or in aquaculture. However, on some occasions, especially for polluted waters and effluents, $SO_4$ may occur in quite large concentrations and the necessity of estimating this ion may be felt.

*Principle*

$SO_4$ ions are widely estimated turbidimetrically after formation of barium sulphate ($BaSO_4$) through reaction of this anion with barium chloride ($BaCl_2$, $2H_2O$) and making the colloid stable at pH 4.8 by adding sodium acetate-acetic acid buffer.

***Requisites***

(a) Apparatus - Colorimeter

(b) Glasswares - Volumetric flasks 1 l, 25 ml pipette etc.

(c) Reagents

(i) Standard $SO_4$ solution: Weigh accurately 0.2719 g of analytical quality potassium sulphate ($K_2SO_4$). Dissolve in DW and make up the total volume to 1 L. This solution corresponds to 50 ppm of sulphur(S) in the form of $SO_4$.

(ii) Sodium Acetate-Acetic acid buffer (pH 4.8): Dissolve 100 g sodium acetate in about 750 ml DW. To this, add 31ml of glacial acetic acid. Make up the volume to 1L.

(iii) Gum acacia : Dissolved 2.5 g gum acacia in 1 L of DW, keep for one night and filter.

(v) Barium chloride ($BaCl_2$) powder: Available in the market. This should be less than 50 mesh size.

*Procedure*

(a) Preparation of standard curve: Take 0.5ml,1.0ml,2.5ml, 5.0ml and 10.0ml of 50 mg $l^{-1}$ standard $SO_4$ solution in each of 25 ml volumetric flask, respectively. Add a few ml of DW, whenever required to make the total volume around 10 ml. Then, add 1 ml of gum acacia and 1 g of $BaCl_2$ powder to each volumetric flask. Make the volume to 25 ml with DW and shake the contents thoroughly. Read the absorbance of the contents of each of the flask colorimetrically at 440 nm. Prepare standard curve for sulphur (S) as $SO_4$, using the concentrations of $SO_4$ in each of the volumetric flasks and their respective absorbance values.

Conc. of S in 25 ml volume

0.5 ml - 1.0 mg $l^{-1}$

1.0 ml - 2.0 mg $l^{-1}$

2.5 ml - 5.0 mg $l^{-1}$

5.0 ml - 10.0 mg $l^{-1}$

10.0 ml - 20.0 mg $l^{-1}$

(b) Estimation of S and $SO_4$

Take 10 ml of water sample in a 25 ml volumetric flask, add 10 ml sodium acetate-acetic acid buffer and mix thoroughly. Then, add 1 ml gum acacia solution and 1 g $BaCl_2$ powder and shake again. Read the absorbance of this flocculated solution in a colorimeter àt 440 nm as stated earlier. Compare the obtained absorbance with the standard curve to get the ppm concentration of S in the solution of volumetric flask.

*Calculations*

(a) Concentration of S in the water sample present in $SO_4$ (mg $l^{-1}$) form = Reading(mg $l^{-1}$)in standard curve $\times$ 25

(b) Concentration of S (mg $l^{-1}$) $\times$ 3 = concentration of $SO_4$ ions (mg $l^{-1}$).

## 2.8. Calcium and Magnesium

Calcium (Ca) and Magnesium (Mg) are the two common cations, which contribute to productivity not only as nutrient elements but also through their influence on hardness as well as alkalinity in water. One old and reliable method for determination of Ca and Mg in water is by complexometric titration using ethylene diamine tetra acetic acid (EDTA).

*Principle*

EDTA forms chelates with a number of polyvalent cations viz Ca, Mg, F, Mn, Zn etc. This behaviour is utilised for determination of Ca by titration with EDTA, using ammonium purpurate ( murexide) as indicator. Similarly, total amount of Ca and Mg can be determined through titration with EDTA after adding Eriochrome black T as indicator. Concentration of Mg in the water samples may then be determined from the difference of values of Ca+ Mg and Ca alone.

*Requisites*

Glass wares, 1 litre vol. flasks, burette, pipette.100ml conical flask etc.

*Reagents*

(i) 0.01 N calcium solution : Dissolve 0.5005 g dried pure calcium carbonate ($CaCO_3$) in about 100 ml of approximately 0.2N HCI (Dilute 5.0 ml concentrated HCl to 250 ml with DW). Boil the solution for some time, cool it down and make up the vol to 1l.

(ii) Ammonium purpurate (Murexide) indicator : Mix thoroughly 0.2 g murexide powder with 40 g potassium sulphate and store in a dark coloured bottle.

(iii) Eriochrome black T indicator: Dissolve 0. 5 g eriochrome black T and 4.5 g hydroxylamine hydrochloride indicator in 100 ml ethyl alcohol. Add 5 ml of 2 per cent sodium cyanide solution to it.

(iv) 4 N sodium hydroxide (NaOH): Dissolve 160 g NaOH in llitre DW. This solution need not be standardized.

(v) Ammonium chloride - ammonium hydroxide ($NH_4Cl$-$NH_4OH$) buffer solution: Mix 67.5 g $NH_4Cl$ and 5.70 ml $NH_4OH$ and dilute to a total volume of 1 litre.

(vi) 0.01 N EDTA solution: Weigh 2.0 g disodium dihydrogen ethylene diamine tetra acetic acid (Di-Na salt of EDTA). Dissolve in 1L of DW. Standardise the solution against standard 0.01 N Ca solution by following the method described below.

The buffer, EDTA and eriochrome black T solutions should be prepared fresh during use for a few days.

*Procedure*

Standardisation of EDTA solution: Take 5 ml 0.01 N Ca solution in a 100 ml conical flask, dilute to about 25 ml with DW. Add 5 m14N NaOH solution and about 25 mg of murexide indicator. Mix the contents well, to get a orange red colouration. Titrate it with 0.01 N EDTA solution against a white background until the colour changes to purple.

The colour change is not very distinct and hence the titration during the end point should be carried out cautiously. It is generally advised to keep an un-titrated sample, after addition of indicator, by the side of the sample being titrated to distinguish the colour change more precisely.

Find out the strength (factor) of EDTA solution by using the following relationship.

Factor = 5/v where V= volume of EDTA solution required.

*Determination of Ca*

Take 5 ml of water sample in a 100 ml conical flask and proceed as in case of standardization process.

*Determination of $Ca^{2+}$+$Mg^{2+}$*

Take 5 ml of water sample in a 100 ml conical flask and dilute it to about 25 ml as before. Add 1 ml of $NH_4$ Cl-$NH_4OH$ buffer and 3-4 drops of eriochrome black-T indicator to get a wine red colour. Titrate against 0.01 N EDTA until the colour changes to blue.

*Calculations*

(a) Concentration of Ca (mg $l^{-1}$) = ml of EDTA used × F × 40

(b) Concentration of Mg (mg $l^{-1}$) =(ml of EDTA used for Ca +Mg - ml of EDTA used for Ca) × F × 24

Where, F=Strength of EDTA solution.

*Comments*

Calcium may also be determined with the help of a flame photometer by following the method as described for potassium and sodium estimation using calcium filter.

## 2.9. Hardness

Hardness of water is caused principally by the elements Ca and Mg and sometimes by iron and aluminium (Theroux et al, 1943). For most practical purposes, however, determination of $Ca^{2+}$+$Mg^{2+}$ usually provides a fair idea of hardness of water and is expressed in terms of calcium carbonate ($CaCO_3$).

*Principle*

For the purpose of determining hardness of water, concentrations of Ca and Mg in water are first determined. The values are then converted to respective equivalents of $CaCO_3$, and added to get total hardness of water.

*Requisites*

Same as sec 2.8, for determination of Ca and Mg.

*Procedure*

Determine concentrations of Ca and Mg in water samples by following the procedure described in Sec 2.2-5

*Calculation*

(a) Observed concentration (mg $l^{-1}$) of Ca as $CaCO_3$ equivalent= mg $l^{-1}$ of Ca in water x Eq. weight of $CaCO_3$ / Eq. weight of Ca= mg $l^{-1}$ of Ca in water x 50.04/ 20.04.

(b) Observed concentration (mg $l^{-1}$) of Mg as $CaCO_3$ equivalent= mg $l^{-1}$

of Mg in water x Eq. weight of $CaCO_3$/Eq. weight of Mg = mg $l^{-1}$ of g in water × 50.04/12.16.

Total hardness of water = (a) + (b) mg $l^{-1}$)as $CaCO_3$.

## *2.10. Potassium*

Potassium is not commonly included in routine analyses of irrigation water. However, a knowledge about the relative occurrence of this cation helps in better understanding of the water quality. Similarly,fish ponds are also usually considered to be fairly well supplied with potassium (Mandal and Chattopadhyay, 1992) and estimation of K has not so far received much attention in productivity assessment programmes of various aquaculture systems. However, with the intensification of fish culture practices and increasing uses of N and P fertilisers, K is gradually becoming the limiting nutrient element and understanding the K status in pond water is becoming an important proposition for judicious nutrient management in aquaculture.

*Principle*

K can be estimated gravimetrically, volumetrically, colorimetrically, potentiometrically or turbidimetrically (Yadav and Khera, 1993). However, in recent years, flame photometric analysis of K has become highly popular. Flame photometry involves emission of typically coloured flames from the basic ions, intensity of which are measured through a galvanometer. It is a rapid method and can be carried out with small amount of water sample.

*Requisites*

(a) Apparatus: Flame photometer with fuel gas supply.

(b) Glass wares : 50 ml beakers. 1 l and 100 ml volumetric flasks.

(c) Reagents : Standard K solution: Weigh 191 mg of AR grade potassium chloride (KCI) and dissolve in DW. Make the volume to 1 litre. This is 100 ppm stock solution of K. Take 10 ml of 100 ppm solution in a 100 ml volumetric flask and make up the volume to get 10 ppm K solution.

*Procedure*

Filter about 20 ml of water sample in a clean 50 mi beaker. Operate the flame photometer by igniting the flame, maintaining proper ratio of fuel gas and compressor air, as per the instructions of the instrument manual. Take some DW in a 50 ml beaker and atomise it through the atomiser. Adjust galvanometer reading to zero with DW. Then, feed the 10 ppm K solution and adjust the galvanometer reading to 100. After adjustment, each unit (one) reading of galvanometer will be equivalent to 0.1 mg $l^{-1}$ of K. Now atomise the unknown water sample into the flame and record the reading.

*Calculation*

Concentration (mg $l^{-1}$) of K in water = Reading x 0.1.

*Comments*

(a) The strength of standard solution ànd adjustment of galvanometer reading may be suitably modified as per the expected concentration of K in water.

(b) During estimation of large numbers of water samples, it is wise to check the galvanometer reading with the standard solution time to time. However, now a days many instruments are having in-built automatic check knob.

### *2.11. Sodium*

Sodium forms an important constituent of analysis for assessment of water quality , especially in saline and alkaline regions. On the other hand, Na is generally not included in regular analyses of pond water. However, in brackishwater ponds, Na occurs in very high concentrations and determination of this element may be necessary.

*Principle*

As stated in case of K, flame photometric analysis is very popular and used for determination of water soluble Na.

*Requisites*

(a) Apparatus: Flame photometer with fuel gas supply

(b) Glass wares : 50 ml beakers, 1 and 100 ml vol. flasks,

(c) Reagents: Standard Na solution: Weight 0.254 g of AR grade dried sodium chloride (NaCl) and dissolve in DW. Make the volume of the solution to 1 litre. This is 100 ppm stock solution for Na. Take 10 ml of 100 ppm solution in a 100 ml vol. flask and make up the volume to 100 mi to get 10 ppm Na solution.

*Procedure*

Filter about 20 ml of water sample in a clean 50 ml beaker. Operate the flame photometer in the same way as stated in sec 2.7 and adjust the reading in such a manner so that each unit (reading) of the instrument becomes equivalent to 0.1 ppm of Na. Now, use the water sample under study and record the reading.

*Calculation*

Concentration (ppm) of Na in water = Reading 0.1

*Comments*

Effect of Na is usually better understood when studied with relation to other cations, The proportion of Na ions in water with relation to total cation concentration, is termed as soluble sodium percentage (SSP).

$$SSP = \frac{\text{Conc. of soluble Na (meq/l)} \times 100}{\text{Total cation conc. (meq/l)}}$$

For practical purposes, total cation concentration includes concentrations of Ca+ Mg + Na +K, expressed in meq/1 (mg $l^{-1}$/ equivalent wt for each cation).

Sodium absorption ratio (SAR) is a term, which is used to express the relative concentration of Na ions over divalent cations viz Ca and Mg.

$$SAR = \frac{\text{Conc. of Na in( meq/l )}}{\text{Conc. (meq/l) of } (Ca^{2+} + Mg^{2+})/2}$$

***2.12. Dissolved inorganic nitrogen***

Nitrogen is an important plant nutrient required both for agriculture and aquaculture . Inorganic forms of N dissolved in water are present predominantly in ammonium ($NH_4$) and nitrate($NO_3$) forms except in highly polluted water, where nitrite ($NO_2$) form may also occur in considerable concentration. Estimation of these two forms of N, therefore, constitutes a measure of availability of N in pond water, while determination of nitrite is often carried out in polluted waters in most cases.

*A. Ammonium and nitrate forms*

*Principle*

To determine the amount if $NH_4$ and $NO_3$ forms of N in water, $NH_4$.-N is first distilled with alkali to release àmmonia ($NH_3$) which is absorbed in an acid.

Then, $NO_3$ is reduced to $NH_4$ with hydrogen ($H_2$) in alkaline solution and the $NH_4$ produced is estimated as before. In the present method, the released $NH_3$ is absorbed in boric acid ($H_3BO_3$) to form $NH_4^-$ borate. This $NH_4^-$ borate is titrated back to original $H_3BO_3$ with a standard acid and the concentration of $NH_4$-N is determined from the amount of standard acid required for this titration.

***Requisites***

(a) Apparatus : Kjeldahl distillation set with condenser.

(b) Glasswares : Kjeldahl flask,250 ml conical flask, beaker, 1litre vol. flask, pipette, burette etc

(c) Reagents

(i) 40 per cent sodium hydroxide (NaOH) solution: Dissolve 400 g NaOH in a 1l beaker using about 750 ml DW, allow the solution to be cooled and make up the volume to 1litrel.

(ii) Davarda's alloy (Cu 50; Al 45 and Zn 5 per cent each): Available in market.

(iii) Mixed indicator: Dissolve 0.5 g bromocresol green and 0.1 g methyl red in 100 ml 95 per cent ethyl alcohol. Adjust the solution to bluish purple mid colour (pH 4.5) with dilute HCl or NaOH.

(iv) 4 per cent $H_3BO_3$ solution : Dissolve 40 g $H_3BO_3$ in about 750 ml DW, add 5 ml of mixed indicator and make up the volume to 1l. This stock solution should be adjusted by gradual addition of dilute HCl or $H_2SO_4$ until the bluish colour changes to pink.

(v) Standard (0.02 N) $H_2SO_4$

*Procedure*

(a) $NH_4^+$ - N

Take 250 ml of filtered water sample in a kjeldahl flask and fit the flask with the distillation set along with condenser. Add 20 ml of 4 per cent $H_3BO_3$ in a 250 ml conical flask and put it below the outlet of the condenser, so that the tip of the outlet dips into the $H_3BO_3$ solution. Add slowly 10 ml of 40 per cent NaOH in the Kjeldahl flask and fix the mouth of the condenser immediately. Carry out distillation till about 40 ml of distillate is accumulated in the receiving flask containing $H_3BO_3$ . Stop heating and take out the conical flask. Titrate the distillate with 0.02 N $H_2SO_4$ till a pinkish colour appears.

(b) $NO_3$—N

Same (residual) water sample after $NH_4^+$estimation can be used for determination if $NO_3^-$ N. Take 20ml 4 percent $H_3BO_3$ in another conical flask and place beneath the condenser tip as before. Add about. 0.5 g Davarda's alloy to the water sàmple, close the mouth of the Kjeldahl flask immediately and continue distillation to near dryness of water sample. Titrate the $NH_4$ borate in the conical flask as stated earlier.

(c) Total Inorganic ($NH_4^+$ + $NO_3^-$) N

In case, separate estimations of $NH_4^+$ and $NH_3^+$ forms not are required and it is required to determine the total dissolved inorganic N ($NH_4^+$ + $NO_3^-$) only, 250 ml of filtered water sample may be straightway distilled after addition of Davarda's alloy and 10 per cent NaOH to almost dryness. The liberated $NH_3$is absorbed in 4 per cent $H_3BO_3$ and determined as stated above.

*Calculations*

N (mg $l^{-1}$) in the form of $NH_4^+$ or $NO_3^-$ or ($NH_4^+$ + $NO_3^-$) = A/V x 280 where A = ml of 0.02 N $H_2SO_4$ required for titration in respective case. V = volume (ml) of water sample used.

*Alternative methods*

$NH_4^+$ and $NO_3^-$ forms of N can also be determined by using colorimetric methods using Nessler's reagent and phenol disulphonic acid, respectively. However, in case of presence of considerable amount of salts in water, both of these methods may suffer from interference.

*Comments*

(a) Neither the volume nor the strength of $H_3BO_3$ need to be known exactly because it is the $NH_4^-$ borate formed that is titrated back to $H_3BO_3$ by $H_2SO_4$ (Jackson, 1971).

(b) To avoid bumping ànd frothing, add a few glass beads and a little amount of liquid paraffin to the water sample.

(c) Nitrite ($NO_2^-$) ions àre usually present in very small amount in ordinary water bodies and they also become reduced and included along with $NO_3^-$. In case such inclusion is to be avoided, 1 ml of aqueous sulfamic acid may be applied to the flask before àddition of Davarda's alloy (Yadav and Kher, 1993).

*B. Nitrite form*

*Principle*

In a strongly acid medium, $HNO_2$reacts with sulphanilamide to form a diazonium compound which reacts quantitatively with N- (1-naphthyl) ethylene diamine d$^i$ hydro chloride to form a strongly coloured azo compound (Golterman *et. al.*, 1978). The intensity of this colour can be measured colorimetrically and the concentration of $NO_2^-$ ion in the sample is determined.

*Requisites*

(a) Apparatus: Colorimeter or spectrophotometer.

(b) Glass wares: 1litre, 100 ml volumetric flasks, pipettes etc.

(c) Reagents

(i) 6 N Hydrochloric acid (HCl) : Dilute AR quality HCl with equal amount of DW to get approximately 6 N HCl.

(ii) 0.2 per cent Sulphanilamide: Dissolve 0.2 g of the chemical in 100 ml DW.

(iii) 0.1 percent N (1-naphthyl) ethylene diamine d$^i$ hydrochloride : Dissolve 0.1 g of the reagent in 100 ml DW. Store in a brown bottle.

(iv) 5 per cent Ammonium sulphamate ($NH_2SO_3$ $NH_4$): Dissolve 5 g of the reagent in 100 ml DW.

(v) 5 N Sodium hydroxide (NaOH):Dissolve 200g NaOH in 1l DW.

(vi) Standard nitrite solution (1000 mg $l^{-1}$) : Dissolve 6.08g of AR quality potassium nitrite ($KNO_2$),dried at $105^0$ C in DW. Add 1 ml 5 N NaOH and dilute to exactly 1 litre. Dilute this solution 50 times to get 20 mg $l^{-1}$ standard solution of $NO_2^-$ -N.

*Procedure*

Take 90 ml of water sample in a 100 ml volumetric flask, add 5 ml sulphanilamide solution and 2 ml 6N HCl and mix thoroughly. Allow the mixture to settle for 3 minutes, add 1 ml of 5 per cent ammonium sulphamate, mix, wait for another 3 minutes and add 1 ml napthylethylene diamine solution. Make up the total volume to 100 ml with DW. Wait for 15 minutes and read the absorbance of the solution at 530 nm against a blank solution.

Prepare a standard curve by using different concentrations of the standard solution and following the same procedure for estimation of $NO_2^-$. Use 2 ml, 5 ml,10ml,20 ml and 50ml of the 20 (mg $l^{-1}$ standard $NO_2^-$ N solution in 100 ml volumetric flasks and proceed as stated earlier). Under final volume of 100 ml, these solutions will have 04, 10, 20, 40 and 100 (mg $l^{-1}$ concentration of $NO_2^-$ N) respectively.

*Calculation*

If observed concentration of $NO_2^-$ N of the water sample, as compared to the standard curve, is X ppm, actual concentration of $NO_2^-$ N in the sample will be X × 1.11 ppm.

*Comments*

$NO_2^-$ is very unstable in nature ànd should be analysed immediately after collection. In case it is not possible to carry out the analysis immediately, the sample may be treated with 2 ml of 6N HCl and 5 ml sulphanilamide solution and preserved for àbout 24 hrs in refrigerator without much problem.

## *2.13. Phosphorus*

Phosphorus (P) is considered as the most critical nutrient element in the maintenance of pond productivity (Thingran, 1975) Owing to its utmost importance to primary fish food organisms and extremely low availability in most the aquatic ecosystems, this element has a special significance in fish culture.

*Principle:*

In natural waters P occurs mostly as phosphate. This form of P is generely estimated colorimetrically after development of phospho- molybdic blue colour (Dickman and Bray, 1940). Stannous chloride ($SnCl_2$) has long been used as the reducing agent for developing the blue colour. But, the colour developed by $SnCl_2$ does not stay long and hence ascorbic acid is commonly used instead of $SnCl_2$ (Murphy and Riley, 1962). The blue colour developed with ascorbic acid is stable for about 2 hrs.

*Requisites:*

(a) Apparatus: Colorimeter or Spectrophotometer.

(b) Glass wares: Vol. flasks 100 ml, 25 ml. pipette *etc.*

(c) Reagents

(i) Standard P solution: Dissolve 0.2195 g anhydrous potassium dihydrogen orthophosphate ($KH_2PO_4$) in about 500ml of DW, add 5 ml conc. $H_2SO_4$ make up the volume to 1 litre. This forms 50 ppm solution of P. 4 ml of this 50 ppm solution is to be diluted to 100 ml to get a 2 ppm solution of P, which is used for making standard curve for P.

(ii) Ammonium molybdate- Potassium antimony tartarate solution : Dissolve 1.0g ammonium molybdate $(NH_4)_6Mo_7O_{24}$ , 4 $H_2O$ and 0.02 g potassium antimony tartarate in about 500 ml DW. Add slowly 15 ml of conc. $H_2SO_4$ to this solution, make up the volume to 1 litre with DW.

(iii) Ascorbic acid-molybdate solution weigh 88 mg ascorbic acid and add to 100 ml of the molybdate-tartarate solution. This solution should be freshly prepared before P estimation.

*Procedure:*

(a) Prepare standard curve with different concentrations of P from the 2 mg $l^{-1}$ solution. Use of 0.25, 0.50, 1.25, 2.5 and 5.0 ml of 2 mg $l^{-1}$ P solution in 25 ml volumetric flasks will result in respective concentrations of 0.02, 0.04, 0.1, 0.2 and 0.4 of 2 mg $l^{-1}$ after making up the final volume.

Take the above mentioned volumes of 2 mg $l^{-1}$ P solution in 5 nos. of 25 ml volumetric flasks and also run a blank in another flask without any P solution. Add some amount of DW in each of the flasks so that the total volume comes up to 10-12 ml. Add 5 ml of ascorbic acid-molybdate mixture to each of the flasks. Make up the volume to 25 ml with DW and wait for 30 minutes for full development of blue colour. Read the optical density in a colorimeter or spectrophotometer at 660 nm wave length after setting O optical density (OD) with the blank solution. Prepare a standard curve by plotting concentrations ( mg $l^{-1}$) of P in X axis and respective optical densities in Y axis.

(b) Take5-10 ml of water samples in 25 ml volumetric flasks depending on expected concentrations of P in water. Also, take a blank flask as before. Add adequate amount of DW to each flask and 5 ml of àscorbic acid-molybdate mixture. Make up the volume to 25 ml and determine the OD of the developed blue colour as before. Put the OD values of unknown water samples on the standard curve to get the concentration of P in the volumetric flasks as stated earlier.

*Calculations*

If the concentration of Đ in 25 ml volumetric flask is X ppm and the initial volume of water sample taken in the flask is V ml,

$$\text{Concentration ( mg l}^{-1}\text{) of P in water sample} = \frac{X \times 25}{V}$$

*Comments*

(a) While estimating P, attention should be paid to keep the glass wares and reagents completely free from P, because very minute concentrations of P may interfere with P estimation. Use of double DW gives better results.

(b) Sometimes $P_2O_5$ is used to denote the concentration of P in water . Approximate relationship between $P_2O_5$ and P is $P \times 2.3 = P_2O_5$

### *2.14.. Dissolved organic matter*

Amount of oxygen utilised during the process of oxidation of organic matter can provide a gross idea about the quantum of dissolved organic matter in water. Potassium permanganate ($KMnO_4$) is commonly used for this oxidation. Hence , the value is sometimes referred as potassium permanganate demand of water also.

*Principle*

When $KMnO_4$ is used for oxidising the soluble organic matter in water, the $MnO_4^-$ ion is reduced to $MnO_2$ form. The change in colour during this process is used to determine the amount of oxygen required to oxidise the organic matter.

*Requisites*

(a) Glass wares: Volumetric flasks (1 *litre*), conical flasks (250 ml), pipette, burette etc.

(b) Reagents

(i) Potassium permanganate ($KMnO_4$) solution : Dissolve 0.4 g $KMnO_4$ in 250 ml DW. Each ml of this solution will be equivalent to 0.1 mg $O_2$.

(ii) Ammonium oxalate ($C_2H_8O_4N_2$): Dissolve 0.888 g $C_2H_8O_4N_2$ in 1 *litre* DW.

(iii) Dilute $H_2SO_4$ : Add slowly and cautiously 100 ml conc.$H_2SO_4$ to 300 ml DW.

*Procedure*

Take 50 ml filtered water sample in a 250 ml conical flask, add 5 ml 1:3 $H_2SO_4$ and shake the flask gently. Add 10 ml of $KMnO_4$ to this acidified water and heat in a water bath. After half an hour, remove the flask from the water bath and add 10 ml ammonium oxalate to it. This will result in disappearance of the existing pink colour of the solution. Now, add to this colourless solution, $KMnO_4$ very slowly till the pink colour just reappears and note the burette reading.

*Calculations*

Amount of $O_2$ ($mgl^{-1}$) required to oxidise the dissolved OM = ml of $KMnO_4$ required X 2.

*Comments*

If the water contains high amount of dissolved organic matter , the pink colour may disappear during the period of heating itself. Under such condition, the procedure should be repeated using higher amount of $KMnO_4$ solution and also equal amount of ammonium oxalate.

### 2.15. *Dissolved Oxógen*

Although not used commonly for irrigation water, dissolved oxygen (DO) is a widely analysed chemical property for fish pond water quality assessment. Estimation of DO is sometimes used for understanding the effluent quality also. Two most popular methods of DO estimation are-i) Winkler's method, and ii) Polarographic technique. Of these, the method of Winkler ( 1888) has remained a widely adopted technique for estimation of DO since long time. However, after development of polarographic technique, this method has rapidly become popular throughout the world owing to its easiness in operation. Considering the simplicity and rapidness of polarographic method, this technology of DO estimation has been discussed here.

*Principle*

Polarographic oxygen analyser consists of a probe comprising of two electrodes , bathed with potassium chloride (KCl) and separated from the water sample by means of a membrane. The electrode produces a current proportional to the tension of oxygen in water and the instrument converts the current to a meter indicating the DO concentration. Operation procedures of different DO analysers are simple . However, they vary in different instruments depending on manufacturer's instructions. Hence, they have not been dealt here.

### *2.16. Free Carbon Di-oxide*

Apart from its important role in photosynthesis of primary fish food organisms, free carbon di-oxide ($CO_2$) has interdependence with pH and bicarbonate-carbonate equilibrium. It also renders some essential nutrient elements in soluble form through formation of carbonic acid. However, if present in higher concentration, it may exert adverse effects on respiration and other physiological functions of biotic lives present in the pond ecosystem.

*Principle*

Water having pH> 8.3 ,does not contain appreciable amount of free $CO_2$ Therefore, phenolphthalein indicator, which has end point àt this pH value, is used for estimating the àmount of $CO_2$ in a water sample for titration with a standard alkali.

$$2NaOH + CO_2 = Na_2\,CO_3 + H_2O$$

*Requisites*

(a) Glass wares: Burette, pipette, volumetric flasks, porcelain basin etc

(b) Reagents :

(i) Phenolphthalein indicator :

Dissolve 0.5 g of phenolphthalein in 50 ml of 95 per cent ethyl alcohol. Make up the volume to 100 ml with recently boiled DW.

(ii) N/44 sodium hydroxide (NaOH):

Prepare a N/10 stock solution of NaOH by dissolving exactly 4.0g AR quality NaOH pellets in 1 *l* DW. Standardise the solution by titrating it against N/10 $H_2SO_4$ using phenolphthalein indicator until a faint pink colour appears. Dilute 100 ml of this N/10 NaOH solution to 440 ml with DW to get N/44 NaOH solution. The prepared standard solution should not be used for a long period and while preparing fresh standard solutions, the stock solution should be standardised every time.

*Procedure*

Take 100 ml of the freshly collected water sample in a white porcelain basin. Add 3-4 drops of phenolphthalein indicator into the water. If the sample turns pink, the pH of water is above 8.3 and free $CO_2$ is not present. If the solution remains colourless after addition of indicator, titrate it with N/44 NaOH with gentle stirring with a glass rod till the colour turns pink.

*Calculation*

Concentration of free $CO_2$ (mg $l^{-1}$) = ml of N/44 NaOH required for titration × 10.

*Comments*

Free $CO_2$ is very labile and hence, water samples for free estimation should be collected with caution so that it does not come in undue contact with atmospheric water. The same càre should be exercised at the time of analysis also so that minimum changes occur in the water sample due to exposure to atmosphere. Boyd (1978) suggested that this analysis should be completed within 2- 3 h of collection.

### *2.17. Biochemical Oxógen Demand*

The biochemical oxygen demand (BOD) is a measure of the amount of oxygen required by micro organisms for decomposing the organic matter (OM) in water under a specific set of conditions . This is estimated particularly when the water is rich in organic materials. Complete decomposition of OM may require a long time and hence a 5 day incubation period has been suggested for most practical purposes (APHA, 1971). The BOD value thus obtained is termed as $BOD_5$

*Principle*

For determination of BOD, an aliquot of the water sample is incubated in dark at $20^0$C for 5 days and the final DO level in this sample is deducted from the initial DO value of water. This gives an estimate of the amount of oxygen required to decompose the OM in the water body during 5 day period under the specified condition. In case of water samples rich in OM, the consumption of oxygen may be so high, that the DO level of water may be completely exhausted. In such cases, suitable dilutions of the water samples are done in order to obtain the BOD value.

*Requisites:*

(a) Apparatus: BOD incubator, DO meter

(b) Glassware: Conical flask, BOD bottle

(c) Reagents: Use the chemicals for DO estimation

*Procedure*

Take about1 *litre* of water sample in a suitable container and shake it thoroughly to increase its DO level. Fill two BOD bottles (narrow neck glass stoppered bottle of usually 250 ml volume) with the water sample. One of these bottles should be black coloured or wrapped with black sheet to avoid entry of light. Incubate this opaque bottle in a BOD incubator at 20°C. Determine the DO of the other bottle (DI) by following the method stated in sec 2.15. Bring out the incubated bottle from the incubator after 5 days and determine the DO of this water sample ($D_5$).

*Calculations*

$$BOD_5\ (mg\ l^{-1}) = DI\ (mg\ l^{-1}) - D_5\ (mg\ l^{-1})$$

*Comments*

In some cases, OM content of the water sample may be so high that there may be absolute depletion of DO in incubated water after 5 days. In such condition , the water sample should be diluted with DW at suitable ratios depending on the quality of the water sample. Use this diluted water as the sample and proceed as stated above. If the water has been diluted X times:

$$BOD_5\ (mg\ l^{-1}) = [DI\ (mg\ l^{-1}) - D_5\ (mg\ l^{-1}) \times X$$

## 2.18. Chemical Oxygen Demand

The chemical oxygen demand (COD) is a measure of the total amount of oxygen, which is required to oxidise all the organic matter in a sample (Boyd, 1978). This property therefore, indicates the organic matter concentrations of pond water and also exhibits close correlation with BOD values as well (Sawyer and McCarty, 1967).

*Principle*

In the measurement of COD, all the organic matter in present a water body are oxidised with the help of a standard oxidizing agent. The amount of oxidising agent consumed during this process indicates the quantity of oxygen required for this purpose and from this COD is determined. In the method presented in this section, potassium dichromate ($K_2Cr_2O_7$) is used for carrying out the oxidation. The excess $K_2Cr_2O_7$ is back titrated by standard ferrous ammonium sulphate ($Fe(NH_4)_2(SO_4)_2$) solution to know the actual amount of standard $K_2Cr_2O_7$ required for oxidation of OM in the water sample. The equivalent requirement of oxygen for oxidising the same amount of OM as has been oxidised by $K_2Cr_2O_7$, is then worked out.

*Requisites*

(a) Apparatus :Reflux system

(b) Glasswares : 250 ml conical flasks, burette, pipette, 1 litre volumetric flasks etc.

(c) Reagents

(i) In Potassium dichromate ($K_2Cr_2O_7$ ): Dissolve 49.036 g dry AR quality $K_2Cr_2O_7$ in about750ml of DW. Dilute it to exactly 1 litre. This serves as stock solution.

(ii) 0.025 N $K_2Cr_2O_7$: Dilute 25 ml of thå stock (1N) solution to exactly 1 litre.

(iii) 0.025 N Ferrous àmmonium sulphate ($Fe(NH_4)_2(SO_4)_2$) : Dissolve 9.8 g of ($Fe(NH_4)_2(SO_4)_2$), 6 $H_2O$ about 500 ml of DW, add about 20 ml of conc $H_2SO_4$, cool the solution, make up the volume to 1 litre. This reagent is easily decomposable and hence should be standardised with $K_2Cr_2O_7$ before use.

(iv) Diphenyl amine indicator : Dissolve 1 g of the indicator in 100 ml of conc., $H_2SO_4$.

(iv) Orthophosphoric acid : Available in market.

(v) Silver sulphate crystals Available in market.

*Procedure*

Take 20 ml of water sample in a conical flask. Add a few glassbeads. Mix exactly 10ml of 0.025 N$K_2Cr_2O_7$ to the water sample and then add slowly 30ml of conc. $H_2SO_4$ if the water is likely to contain large amount or chloride e.g, in case of estuarine waters add a few crystals of silver sulphate to the flask to overcome the interference of these ions. Attach the flask to a reflux condenser and reflux for 2 hrs over a heater. After refluxing, wash the inside residues of the condenser into the flask, dilute with àbout 75-80 ml DW. Add 10 ml of acid to prevent interference of ferrous ions during titration. Add about 1 ml of diphenyl amine indicator to develop a blue colour in the solution and titrate with standard $Fe(NH_4)_2(SO_4)_2$, until the blue flushes to green. Run a blank in exactly the same procedure with all the reagents.

*Calculation*

Boyd (1978) suggested a simple calculation for determinationof COD.

$$\text{COD (ppm)} = \frac{X\,(N) \times 8000}{V}$$

Where X = ml of Fe $(NH_4)_2$ $(SO_4)_2$ required in blank - ml of Fe $(NH_4)_2$ $(SO_4)_2$ required in sample titration.

N-Normality of Fe $(NH_4)_2$ $(SO_4)_2$ solution.

V-sample volume.

*Comments*

In some cases, the COD of water may be so high that 10ml 0.025N$K_2Cr_2O_7$ may not be sufficient to oxidize all of the OM present in it. In such cases, higher amount of conc. of $K_2Cr_2O_7$ may be used.

### *2.19. Chlorine*

Free chlorine remains in very small concentrations in pond water. However, the amount may be higher under contaminated or polluted conditions and need to be estimated.

*Principle*

Free chlorine reacts with N-N diethyl-P-phenylene diamine(DPD) in presence of excess iodide to produce a red colour. This colour disappears when the chlorine is reduced to chloride by ferrous ammonium sulphate (Santhanam *et al.*, 1989). This reaction is utilised to determine the amount of free chlorine in pond water.

$$\underset{\text{Red}}{\text{Chlorine — DPD}} + Fe^{2+} \longrightarrow \underset{\text{colourless}}{Cl^- + Fe^{3+} + \text{DPD}}$$

*Requisites*

(a) Glasswares:250 ml conical flasks, pipettes, 250 ml, 50ml,1 litre volumetric flasks.

(b) Reagents

(i) DPD indicator : Dissolve 250mg N-N-diethyl-P-Phenylene diamine (DPD)-oxalate in about 150 ml DW. Add 2 ml of 25 per cent $H_2S0_4$ (1:3 conc. $H_2S0_4$ : DW). Make the volume to 250 ml with DW. This indicator should not be used for prolonged period.

(i) Buffer solution: Dissolve 12 g disodium hydrogen phosphate ($KH_2PO_4$) in about 250 ml DW and add to the solution 10 mg mercuric chloride ($mgCl_2$). Dissolve 0.4 g ethylene diamine tetra acetic acid (EDTA) salt in 50 ml DW in a separate flask and add to the former mixture. Mix the contents and make up the volume to 1 litre with DW.

(iii) Potassium iodide crystals : Available in market.

(iv) 1 N ferrous ammonium sulphate (Fe $(NH_4)_2(SO_4)_2$): Dissolve 392.2g of Fe $(NH_4)_2(SO_4)_2$ ,6 $H_2O$ in 800 ml of DW, add 20ml conc. $H_2SO_4$ and dilute to 1 litre with DW.

*Procedure*

Mix 5 ml each of the buffer solution and the DPD indicator into a 250 ml conical flask. Add to this flask 100 ml of the water sample and 1g of KP crystals. Mix thoroughly. Titrate the solution with 1N Fe $(NH_4)_2(SO_4)_2$, until the red colour just disappears.

*Calculations*

Conc. of Cl (ppm) = Titre value (ml) × 354.5

*Levels of $Cl_2$ in pond water*

Free Cl is extremely toxic to fish population. White (1955)observed only 0.028 to 0.08 ppm of $Cl_2$ to kill 100 per cent of fish population with varied species. Considering these facts, Boyd (1978) suggested that occurrence of even measurable concentrations of Cl in pond water should not be considered even safe for holding fishes.

### 2.20. Boron

Presence of boron in irrigation water in slightly higher concentration may appear to be toxic, and hence an understanding of the concentration of this element in such water sometimes becomes vital. Analysis of B forms an important component of irrigation quality assessment in saline and alkaline soil zones, where toxic occurrence of B frequently poses a threat to agricultural crops.

*Principle*

Several methods are available for estimation of boron. Among those, colorimetric methods are generally adopted for their easiness and accuracy. Of these, again, analytical method using azomethane –H reagent is popular now a days. Over a concentration range of 0.5 to 10 ug $I^{-1}$, azomethane –H solution forms a stable complex with $H_3BO_3$ at pH 5.1. The intensity of the colour developed during this reaction depends on the concentration of B and can be measured colorimetrically. The method is simple and can be performed under wide range of B concentrations.

*Requisites*

i) Equipment: Colorimeter/Spectrophotometer

ii) 25ml volumetric flasks

iii) Reagents:

a) Buffer solution: Dissolve 15g Di-Na salt of EDTA and 250 g $NH_4$-acetate in about 400 ml DW. Pour 125 ml of glacial acetic acid to the solution, mix well and make up the volume to 1 litre.

b) Azomethine H reagent: Dissolve 0.45 g azomethane H in 100 ml solution of 1% L-ascorbic acid .

c) Standard B (100 mg $l^{-1}$) solution: Dissolve 0.57 g $H_3BO_3$ (AR quality) in 1 litre B free double distilled water.

*Procedure*

Preparation of standard curve: Prepare a 2 mg $l^{-1}$ standard solution of B by taking 2 ml of the 100 mg $l^{-1}$ stock solution into a 100 ml volumetric flask and make up the volume with double distilled water. Prepare standard curve with different concentrations of B from the 2 mg $l^{-1}$ solution. Use 0.25, 0.50,1.25, 2.5 and 5.0 ml of 2 mg $l^{-1}$ B solution in 25 ml volumetric flasks for respective concentrations of 0.02,0.04,0.1,0.2 and 0.4 mg $l^{-1}$ after making up the final volume. Also keep a blank flask without any B solution.

Take the above mentioned volumes of 2 mg $l^{-1}$ solution B solution in 5 nos. of 25 ml volumetric flasks and also run a blank in another flask without ány B. Add some amount of DW in each of the flasks, so that the total volume comes up to 10-12 ml. Add 2 ml buffer solution to each of the flasks, stir well and add 2 ml of azomethane H solution. Mix the contents well and make up the volume to 25 ml with DW and wait for 30 minutes for full development of a pink colour. Read the optical density (OD) with a spectrophotometer at 420 nm wave length. Prepare a standard curve by plotting concentrations (mg $l^{-1}$) of B in Õ axis and respective optical densities in Y axis.

Determination of B: Take 5-10 ml of water sample in 25 ml volumetric flasks depending on expected concentrations of B in waters. Also take a blank flask as before. Add adequate amount of DW to each flask and proceed as before. Determine the OD of the developed colour. Put the OD values of unknown water sample on the standard curve to get the concentration of B in the volumetric flasks as stated earlier.

*Calculations*

If the concentration of B in 25 ml volumetric flask fits to X mg $I^{-1}$ in the standard curve and the initial volume of water sample taken in the flask is V ml,

$$\text{Concentration of B in water sample ( mg l}^{-1}\text{ )} = \frac{X \times 25}{V}$$

*Comments:*

While estimating B, the glass wares and reagents should be completely free from B, because very small concentrations of B may interfere with the estimation.

## 3. Water Quality Critera

For characterizing any water for the purpose of understanding its quality and determining suitable management options, it is essential to fit the observed value of any water parameter to a suitable category. Therefore, any water quality assessment programme can not be successful without a clear knowledge about the criterion of the relevant properties. Different standards of water quality are available for varying uses of water, which may differ from country to country also depending mostly on variations in economic, social and health conditions. In this section, some common water quality standards for two major uses of water in agriculture viz. irrigation and aquaculture have been discussed with special reference to Indian condition.

### 3.1. *Irrigation*

The major characteristics of an irrigation water, that appear to be the most important in determining its quality have been described to be :i) total concentrations of soluble salts (electrical conductivity); ii) relative proportion of sodium (sodium absorption ratio); iii) concentrations of boron or other elements that may be toxic ; and iv) carbonate and bicarbonate concentrations related to the concentration of calcium and magnesium (residual sodium carbonate) by Anon (1968).

Paliwal and Yadav (1976) based on the extensive trials conducted on the suitability of irrigation water for semi-tolerant and tolerant crops in different soil types suggested the upper permissible limits of EC,SAR,RSC, and B for soils (Table 2).

**Table 2 :** Suitabilty of irrigation water in different soils for semi-tolerant and tolerant crops

| Texture | EC ($dSm^{-1}$) | | SAR | | RSC (me/litre) | | B ($\mu g\ ml^{-1}$) | |
|---|---|---|---|---|---|---|---|---|
| | ST | T | ST | T | ST | T | ST | T |
| >30% clay | 1.5 | 2.0 | 10 | 15 | 2 | 3 | 2 | 3 |
| 20-30% clay | 4.0 | 6.0 | 15 | 20 | 3 | 4 | 2 | 3 |
| 10-20% clay | 6.0 | 8.0 | 20 | 25 | 4 | 5 | 2 | 3 |
| < 10% clay | 8.0 | 10.0 | 25 | 30 | 5 | 6 | 3 | 4 |

Minhas and Gupta (1992), while grouping the quality of irrigation water under Indian condition, suggested the following classifications (Table 3).

**Table 3 :** Classification of water quality under Indian conditions.

| Water quality | Electrical Conductivity ($dSm^{-1}$) | Sodium Absorption Ratio | Residual Sodium Carbonate ($meq\ l^{-1}$) |
|---|---|---|---|
| **A. Good water** | <2 | <10 | <2.5 |
| **B. Saline water** | | | |
| i) Marginally saline | 2-4 | <10 | <2.5 |
| ii) Saline | >4 | <10 | <2.5 |
| iii) High-SAR saline | >4 | >10 | <2.5 |
| **C. Alkali water** | | | |
| i) Marginally alkali | <4 | <10 | 2.5-4.0 |
| ii) Alkali | <4 | <10 | >4.0 |
| iii) High SAR alkali | Variable | >10 | >4.0 |

In addition, toxic waters may exhibit excess ions such as nitrate, boron, fluoride, chloride, selenium, cadmium, lead, arsenic *etc* and also some organic/inorganic compounds making the water less suitable for agricultural use. Of these, high concentrations of boron has been considered as a problem for irrigating a large part of agriculturally productive regions of India. Yadav and Khera suggested the following standards of boron in irrigation water for using in agriculture (Table 4)

**Table 4 :** Irrigation water quality rating with respect to boron

| Toxicity | Concentration ($mg\ l^{-1}$) |
|---|---|
| High | <1.0 |
| Medium | 1.0-2.0 |
| High | 2.0-4.0 |
| Very High | >4.0 |

Standards of several other commonly occurring properties of water, which may influence the quality of irrigation water have been regularized through ISI Standards and have been presented below (Table 5).

**Table 5 :** Standards of effluent water used for irrigation

| Property | Thresh hold level (mg $l^{-1}$) |
|---|---|
| TDS | 2100 |
| TSS | 200 |
| BOD5 | 100 |
| COD | 500 |

### *3.2. Aquaculture*

With the changes in objectives and perspectives of water use, standards of water quality become different for aquaculture and some properties like DO, $CO_2$ *etc.* occupy premium position as compared to irrigation water. Water quality parameters of some of the properties relevant to aquaculture have been discussed here.

**A.** In a recent review on pond soil and water management, Adhikari (2011) suggested the following criterion for evaluating some water quality parameters in freshwater aquaculture.

i) **D.O.:** More than 5.0 mg $l^{-1}$ DO is considered favourable for aquaculture, while 1.0-5.0 mg $l^{-1}$ may have sub-lethal effects on growth, feed conversion and tolerance to diseases .

ii) **Free $CO_2$**: 12-50 mg $l^{-1}$ exert adverse effects including respiratory stress, while prolonged exposure to more than 50 mg $l^{-1}$ concentration is likely to be lethal to many fish species.

iii) **pH**: Slightly acidic to moderately alkaline pH range is considered to be optimum. Growth may be slow below pH 6.0 and above 9.0, while stress condition is likely to occur below 4.0 and above 11.0.

iv) **Total Alkalinity**: 75-300 mg $l^{-1}$ is considered to be ideal for fish culture and deviation from this range is likely to result in stress of fishes.

v) **Chlorine**: As low as 0.1 mg $l^{-1}$ can affect fish life.

vi) **Ammonia**: 0.05-0.4 mg $l^{-1}$ concentration is sub-lethal and below and above this range is safe and lethal respectively.

vii) **Nitrite**: Up to 0.02-1.0 mg $l^{-1}$ may be sub-lethal and above this will be lethal.

**B.** In addition, Banerjea (1967), while working on soil and water quality of large numbers of freshwater fish ponds of India, suggested the following criteria for two major nutrients.

i) **Water soluble Nitrogen:** Concentrations below 0.1mg $l^{-1}$, more than 0.1 to 0.2 mg $l^{-1}$ and above 0.2 mg $l^{-1}$ indicators of low, medium and high productivity, respectively.

ii) **Water soluble Phosphorus:** Less than 0.5 mg $l^{-1}$ dissolved phosphate ($P_2O_5$) is considered to be low for a productive fish pond, while in the range of 0.5-2.0 mg $l^{-1}$ medium to high production is expected and highly productive ponds generally exhibit $P_2O_5$ concentration >2.0 mg $l^{-1}$.

iii) **Water soluble Potassium:** Practically no information on criteria of water soluble potassium is available in India. However, Boyd (1984) opined that 1.3mg $l^{-1}$ may be considered to be the critical level of K in fish pond water.

**C.** **$BOD_5$:** Chattopadhyay *et al.* (1988) observed the net primary productivity values of water under $BOD_5$ ranges of 10-20 and >20-30 mg $l^{-1}$ to be significantly higher than the values below 10mg $l^{-1}$. However , late night DO values under >20-30 mg $l^{-1}$ $BOD_5$ level reduced drastically to very low ranges. Considering this behavior, the workers suggested a $BOD_5$ range of >10-20 mg$l^{-1}$ to be conducive for sustaining the productivity of fish culture.

## References

Adhikari, S.P., 2011. Soil and water quality management in aquaculture. In: Handbook of Fisheries and Aquaculture. New Delhi. pp591-622.

Anon, 1968. Diagnosis and Improvement of Saline and Alkaline Soils. USDA Handbook No.60.pp160.

APHA, 1971. Standard Methods for the Examination of Water and Wastewater.Washington DC pp 874.

Banerjea, S. M., 1967. Water quality and soil condition of fish ponds in some states of India in relation to fish production. *Indian J. Fish.* 14: 115-144.

Boyd, C.E., 1978. Water Quality in Warmwater fish ponds. Agril. Expt.Stn. Auburn Univ. pp 359.

Chattopadhyay, G.N.; Saha, P.K. ; Ghosh, A. and Karmakar, H.C., 1988. A study on BOD levels for fish culture in wastewater ponds. *Biol. Wastes.* 25 : 79-83.

Chattopadhyay, G.N., 1998. Chemical Analysis of Fish Pond Soil and Water. Daya Publ. House.pp 132.

Dickman, S.R. and Bray, R.H., 1940. Colorimetric determination of phosphate Ind.Eng, Chem. Anal. 12: 665-668.

Golterman, H.L. ; Clymo, R.S. and Ohnstad M.A. M., 1978. Method for Physical and Chemical Analysts of Fresh Waters. Blackwell Scientific Publicatons London pp. 211.

Gupta, S.K., 2010. Management of Alkali Water. Tech. Bull. Central Soil Salinity Research Institute. Karnal, India.pp 62.

Hickling, C.F., 1971. Fish culture. Faber and Faber. London PP 225.

Jackson, M.L., 1971. Soil Chemical Analysis. Prentice Hall of India Pvt. Ltd. New Delhi pp. 498.

Jhingran, V.G, 1975. Fish and Fisheries of India. Hindustan Publ. Corp. New Delhi pp. 954.

Mandal, L.N. and Chatîpadhyay, G.N. (1992). Nutrient management in aquaculture, In: HLS Tandon (Ed.) Non-Traditional Sectors for Fertiliser Use FDCO, New Delhi, India. pp 1-17.

Minhas and Gupta, R.K. 1992. Quality of Irrigation Water- Assessment and Management. ICAR Pub., New Delhi.pp 123

Murphy, J. E.and Riley, J.Đ., 1962. A modified single solution method for the determination of Phosphate in natural waters. *Analytica Chim. Acta.* 27: 31.

Paliwal, K.V. and Yadav, B.R., 1976. Irrigation water quality and crop production in Delhi territory. Tech. Bull.no. 9. IARI, New Delhi.pp166

Santhanam, R.; Velayutham, P. and Jegatheesan, G, 1989. A Manual for Freshwater Ecology, Daya Publishing House, Delhi. pp 134.

Sawyer, C.N. and McCarty, P.L., 1967. Chemistry for Sanitary Engineers. McGraw-Hill Book Co. New York, pp 518.

Tandon, H.L.S., 1993. Nutrients in soils, plants, waters, fertilisers and organic manures. In : H.L.S.Tandon, . (Ed).Methods of Analysis of Soils, Plants, Waters and Fertilisers. Fertiliser Development and Consultation Organisation. New Delhi.pp1-12

Theroux, F.R.; Eldridge, E.F. and Mallman, W. LeRoy, 1943. Laboratory Manual for Chemical and Bacterial Analysis of Water and Sewage. McGraw Hill Book Co. Inc. London,pp 274.

White, C.E. Jr. 1965. Chlorine: It's Toxicity to Gold fish Fathead Minnows, Golden Shiners and Bluegills and It's Removal from water. Ms thesis, Alabama Poly Inst. Auburn pp 58.

Winkler, L.W., 1888. The Determination of dissolved oxygen in water. *Berline Deutsch Chem Gescllsch* 21: 2843.

Yadav, B.R. and Khera, M.S., 1993. Analysis of irrigation waters. In: HLS Tandon (Ed.) Methods of Analysis of Soils, Plants, Waters and Fertilisers. FDCO. New Delhi pp 144.

❑❑❑

CHAPTER - 7

# Environmental Monitoring

**Dipak Kumar Gupta**

**Earth provides enough to satisfy every man's need but not every man's greed.**

*—Mahatma Gandhi*

The term 'environment' is derived from French word 'environner', which literary means 'to surround' or 'to encircle' i.e. everything either living (plants, animals, microbes) or nonliving (soil, water, air, temperature, humidity etc.) surrounding us makes our environment. The universe is full of uncountable number of stars, planet and satellites, which has specific type of environment but earth is the only known planet in the universe that contains life supporting environment. The earth environment is a collective term for the nature or ecosystems, that we use and benefit from on a local, national and global scale. The earth ecosystems provide number of benefits called **ecosystem services,** free of cost. These services include production of food, fiber, fuel, timber, oxygen, medicine and honey; purification of water and air; waste decomposing and detoxification; flood regulation; carbon sequestration and climate regulation; diseases and pest control; soil formation and nutrient cycling and pollination etc. These services are vital for human beings and other living organisms on the earth to sustain and can be regularly obtained on sustainable basis, if all the components of environment remain in harmony with each other. Any shift from the natural balance of environmental components may have harmful impact on earth ecosystem and living organisms (Fig. 1).

There is a strong link between state of environment and agricultural production as well. Crops require good state of environment for proper growth, development and production, while several agricultural activities have adverse impact on environment. For crops, the state of the environment directly influences soil nutrient availability, water (ground and surface water for irrigation), climate and weather (rainfall and growth season), availability of insects for pollination, and last but not the least, the abundance and effects of certain pests, such as pathogens, insects and weeds, which have major impact on crops worldwide. Without these environmental services, there would be no production. Agriculture is critical to the maintenance of the huge human populations now on earth. It feeds and provides raw materials for about 7 billion people of the world and has challenge to feed expected population of 9 billion by 2050. Towards 2050, rising population and income are expected to call for 70 per cent more food production globally, and up to 100 per cent more in developing countries, relative to 2009 levels (SOLAW, 2011). Ecosystem services enhance agro-ecosystem resilience and sustain agricultural productivity. Thus, promoting the healthy functioning of ecosystems, ensures the sustainability of agriculture as it intensifies to meet the growing demands for food production.

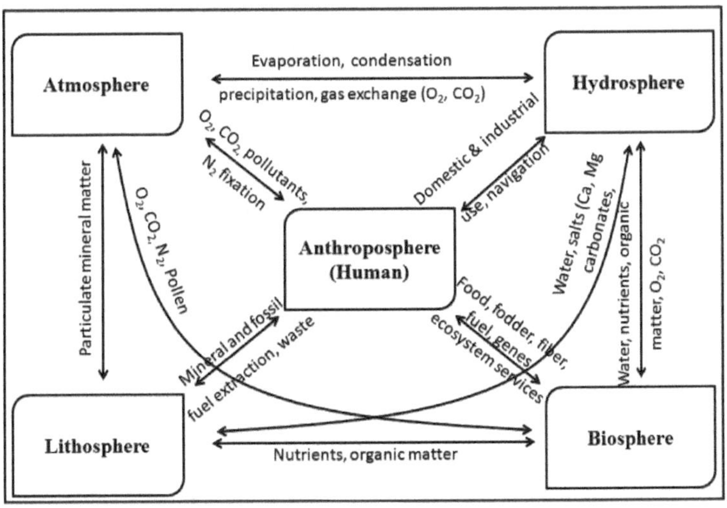


**Fig. 1 :** Interaction of different components of environment and human

However rising human population and its developmental activities like agriculture, industry, transportation, construction, manufacturing, mining etc. has led to severe environmental degradation not only at local and regional scale, but also on global scale (Fig. 2, Table 1).

Out of the several human activities, agriculture is one of the major activities, which have several harmful impacts on environment. Increasing evidence suggest that unsustainable agricultural practices are one of major causes of deforestation, pollution and biodiversity loss and becoming threat to future generation. Agriculture is now also a major cause of global warming and climate change by contributing 13.5% global greenhouse gas (GHG) emissions (IPCC, 2007). One side, agriculture has negative impact on environment another side, environmental degradation will have negative impact on agriculture as well. Rising environmental degradation and depleting natural resources, like land and water is already affecting quantity and quality of agriculture production. At the same time, climate change brings an increasing risk and unpredictability for farmers from warming and related aridity and growing incidence of extreme weather events. At present, agriculture uses 11 per cent of the world's land surface and 70 per cent of all water withdrawn from aquifers, streams and lakes for crop production. In future, there will be very limited scope to increase land and water allocation under agriculture. Land degradation, widespread water scarcity and climate change is posing new threats to agricultural sustainability.

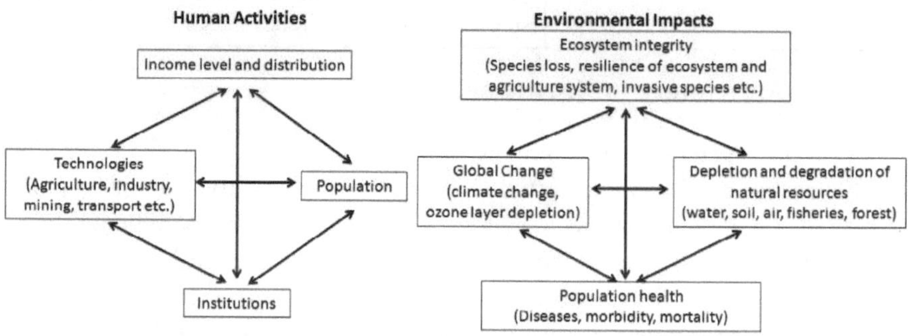


**Fig. 2** : Human activities and its impact on environment

**Table 1** : Global environmental issues and their impacts

| Global change issues | Impacts |
|---|---|
| 1. Global warming and climate change | Heat-trapping gases (Greenhouse gases) emitted by fossil fuel burning is increasing the earth's surface temperature, with numerous potentially negative impacts on ecosystems and human welfare |
| 2. Stratospheric ozone destruction | Certain long lived gases, like chlorofluorocarbons (CFCs) injected into the atmosphere from aerosol spray cans and air conditioners have lowered ozone levels, allowing damaging ultraviolet light to penetrate to the earth's surface. |
| 3. Troposphere ozone pollution | Ozone produced at ground level from pollutants (NOx and hydrocarbons) emitted by vehicles, power plants, and industrial processes causes lung disease in humans and damages plants. |
| 4. Atmospheric traffic | Jetliners produce contrails at 8–13 km altitude that introduce aerosol pollutants into the upper atmosphere, affecting ozone and greenhouse gas levels |
| 5. Deforestation | Tropical rainforests are being cleared for agriculture and settlement around the globe, with negative effects on biodiversity. Forest clearing also increases greenhouse |

| | | |
|---|---|---|
| | | warming and increased erosion |
| 6. | Desertification | Over 70% of the world's semi-arid zone has been moderately to severely damaged by overgrazing and unsustainable agricultural practices. |
| 7. | Coral reef destruction | Coral reefs are thought to be home to 24% of all marine life, yet 70% of the earth's coral reefs could be lost over the next few decades. |
| 8. | Species extinction or biodiversity loss | Poaching, habitat destruction, spread of exotic species, pollution, and global warming are driving many species to extinction. Over 19,000 plant species and 5,000 animal species are classified as endangered |

## Impact of Agriculture on Environment

Human impact on environment started with advent of agriculture about 10,000 years ago, when man started clearing forest for cultivation of crop. But, the rate of forest clearing was very low with sufficient time to regenerate and agriculture was totally dependent on natural inputs and harmful impact was not a matter of concern. 1960 onward, world has witnessed green revolution, which was possible due to intensive agriculture practices like excessive tillage, irrigation, use of chemicals and high yielding varieties. One side this intensive agriculture has led to 3-4 fold increase in food grain production worldwide, other side has also created severe environmental impact. The environmental impact of agriculture is enormous and it has caused environmental degradation at all scale ranging from local to global scale due to several agricultural activities (Fig. 3). These problems are made worse by misdirected subsidies, low public and political awareness of the importance of environment, and weak environmental legislation.

**Impact of Agriculture on Environment**

| Land clearing and tillage | Irrigation | Fertilizer | Pesticide | Mechanical operation & energy | Residue burning | Animal |
|---|---|---|---|---|---|---|
| •Deforestation<br>•Soil erosion<br>•Soil compaction<br>•Air pollution (SPM)<br>•Loss of SOM<br>•Loss of biodiversity | •Ground water depletion<br>•Water logging<br>•Soil salinization<br>•Sea water intrusion<br>•$CH_4$ and $N_2O$ emission<br>•Pathogen and contaminants from waste water | •Nitrate leaching to ground water<br>•$N_2O$ & $NH_3$ emission<br>•Run off of P and N to surface water<br>•Heavy metal contamination (Cd)<br>•Soil acidification | •Surface and ground water pollution<br>•Non target effect: biodiversity loss<br>•Heavy metal contamination (As, Hg, Pb) | •Emission of $CO_2$<br>•Air pollution (NOx, SOx, SPM)<br>•Noise pollution | •Air pollution (NOx, SOx, SPM)<br>•Nutrient loss<br>•Loss of soil biodiversity | •$CH_4$ emission<br>•Surface and groundwater pollution<br>•Desertification |

Land Degradation; Soil, Water, Air & Noise Pollution; Ozone Layer Depletion; Global Warming & Climate Change; Loss of Biodiversity; Poor Human Health

**Fig. 3** : Impact of agricultural activities on environment

## 1. Impact on Soil

Soil is a medium for plant growth and habitat for millions of micro flora and fauna. The agricultural productivity is highly dependent on soil quality, its stability and resilience. The term **soil quality** refers to the soil's capacity to perform its three principal functions e.g. economic productivity, environment regulation, and aesthetic and cultural values. **Soil stability** refers to the susceptibility of soil to change under natural or anthropogenic perturbations. In comparison, **soil resilience** refers to soil's ability to restore its life support processes after being stressed. But, several human activities are deteriorating the soil quality and reducing its stability and resilience capacity. Any undesirable change in physical, chemical and biological characteristics of soil is called **soil pollution**.

Among agricultural activities, clearance of forest for agricultural land; tillage for seed bed preparation; application of fertilizer and pesticide and animal rearing has several harmful impact on the soil (Table 2). The process of deforestation, tillage and open grazing remove soil surface vegetation and expose soil directly to air, rain and wind making soil susceptible to loss organic matter, erosion and desertification. In the major rice-wheat regions of north-western India, **soil organic C** content has decreased from 0.5% in 1960s to 0.2% at present due to excessive tillage. Exposed soil aggregates get easily broken by the energy of raindrops and wind leading soil erosion. On an average, Indian soil loss rate is about16 tons/ha/year, much higher than acceptable rate of erosion at11 tons/ha/year. This adds up to 5334 million tons of soil and 8.2 million tons of nutrients loss every year. Overuse of machinery, intensive cropping, and short crop rotations, intensive grazing and inappropriate soil management leads to **soil compaction**. It is exacerbated by low soil organic matter content and use of tillage or grazing at high soil moisture content. Soil compaction can impair water infiltration into soil, crop emergence, root penetration and crop nutrient and water uptake, all of which result in depressed crop yield. Seasonal cycles of puddling (wet tillage) and drying, over the long term, also lead to the formation of hardpans in paddy soils.

**Table 2** : Physical, chemical and biological process of land and soil degradation

| Physical process | Chemical process | Biological processes |
|---|---|---|
| • Desertification | • Water and wind erosion | • Soil compaction and crusting |
| • Water logging | • Soil salinization | • Soil acidification |
| • Pollution | • Nutrient mining | • Soil organic matter depletion |
| • Denudation of vegetation | • Impairment soil microbial activities | |

Water logging, soil salinity and alkalization, soil acidity, nutrient deficiency, heavy metal and pesticide contamination are other soil related problems associated with agricultural activities. **Water logging** results primarily from inadequate drainage and over irrigation and from seepage of irrigation canal and reservoir. About 3.1 Mha of India's agricultural land is waterlogged and is most serious in Haryana, Punjab, West Bengal, Andhra Pradesh, and Maharashtra. In India, major reservoir like Bhakra canal, Indira Gandhi Nahar Project has led to wide spread water logging in adjoining area. Natural **soil salinity** occurs in hot arid and semi-arid climates with scarce annual rainfall and high evaporation however, soil salinity can occur either due to use of poor quality water like municipal or industrial wastewater for irrigation or rising level of saline ground water due to excessive irrigation combined with inadequate leaching or water logging. Salt-affected soils occupy, on a global basis, 952.2 million ha of land. These soils constitute nearly 7% of the total land area or nearly 33% of the potential agricultural land area of the world. Alkalization, the build- up of sodium in soils, is a particularly detrimental form of salinization, which is difficult to rectify. About 5.95 Mha of India's land is affected by salinity and alkalinity. It is a serious problem in Uttar Pradesh and Gujarat. **Acid soils** occur naturally but agricultural soils can also become acidic due to the continuous additions of large amounts of acid-forming fertilizers, such as ammonia and urea. For example, one mole-equivalent weight of ammonium ($NH_4$) can produce two mole-equivalent of H after it is fully oxidized to nitrate ($NO_3^{-1}$) in the soil environment. In India, about 25 Mha land is affected by soil acidity and use of acidic fertilizer may worse the condition. Most of the Indian farmer applies only nitrogenous and phosphatic fertilizer to their field while only very few farmers apply micronutrients. This unbalanced used of fertilizer has also led to widespread **nutrient deficiency** of micronutrients like Zn, B, Mo etc. in India and other countries. Wastewater, some fertilizer and pesticide contain **heavy metals**. These heavy metals are toxic to human and other animals. Fertilizer mainly phosphate fertilizer contains a majority of the heavy metals like Cd, Hg, As, Pb, Cu, Ni, and Cu; natural radionuclide like $^{238}U$, $^{232}Th$, and $^{210}Po$. Continuous application may cause accumulation of these heavy metals in soil and plants. Plants absorb heavy metals through the soil and thus heavy metal can enter the food chain. The famous incidences of "**itai-itai**" and "**minamata**" diseases due to Cd and Hg toxicity, respectively are the examples of potential threat of heavy metal pollution. In Japan, excessive use of sulphate containing fertilizers in paddy soils resulted in arsenic (As) toxicity, which was attributed to the displacement of arsenate ions from soil particles by sulphate ions.

## 2. Impact on Surface and Ground Water

Water is vital natural resource on earth and immensely important for sustaining life on earth. In spite of plenty of water on earth (2/3$^{rd}$ of earth surface), water is one of the most scarce resource. Out of total water on earth, only 2.5% water is in usable form i.e. fresh water and 97.5% is saltwater, present in ocean and unsuitable for drinking, irrigation and other economic activities. Out of total fresh water, only 1% is easily accessible for economic use as surface water primarily in lakes and rivers, while 79% and 20% is stored in **cryosphere** (polar ice caps and mountain glaciers) and ground water aquifer, respectively which is not easily accessible. Rising population and their increased demand is causing shortage of already scarce fresh water worldwide. In India, the per capita availability of water has been reduced from about 5,177 cubic meters in 1951 to the estimated level of 1,820 cubic meters in 2001 and likely to drop below1, 000 cubic metres by the year 2050.

Agriculture is considered as one of the major causes of shortage of fresh water. The excessive utilization and pollution of available fresh water by agricultural activities, are two most prominent impact, of agriculture on water sources. Among different economic sectors, agriculture is largest consumer of fresh water. It consumes about 70% of planet's accessible fresh water at global level, while in many developing countries like India, it also reaches up to 95%. This is mainly due to increase in extensive irrigation agriculture system. The area under irrigation is increasing worldwide and had increased by about 75% from year 1990. India has second highest total net area under irritation next to China and account for about 25% of the world's irrigation. The irrigation water utilization efficiency in agriculture is very low. About 60% of applied irrigation water in agricultural field get wasted due to faulty irrigation practices like flood irrigation method, seepage and evaporation from irrigation canal.

In many regions, ground water is becoming major source of irrigation and more groundwater is being pumped than is naturally being replenished. As a result, the **groundwater levels** are dropping. In India, tubewell irrigation is increasing very fast mainly in north-western part i.e. Punjab, Haryana western U.P. and Rajasthan mainly due to subsidised rate of electricity and practice of water intensive rice-wheat system. The water table in vast portions of the region, has already reached a level of more than 15-20 meters below the ground surface and is declining at the rate of ranging from 20-100 cm per year in different part of Punjab, Haryana and Rajasthan.

The inefficient use of fertilizer and pesticide and mismanagement of animal waste are major cause of surface and ground water pollution. Fertilizer and pesticide use efficiency in agriculture is quite low. Even with good management practices, efficiency of N, P, K fertilizer and pesticide use seldom exceeds 40%, 20%, 50% and 15%, respectively. Rest of fertilizer and pesticides get wasted and find its place in surface and ground water and atmosphere. When animal waste is over-applied to land and exceeds the capacity of soil and crops to assimilate its nutrients, it becomes a pollutant and can contaminate water supplies and emit harmful gases into the atmosphere. Runoff and leaching from agricultural field and animal waste dumping site causes N and P **nutrient pollution** to both surface and ground water. One of the most important negative effects of nutrient pollution to surface water is **cultural eutrophication.** Eutrophication is the process of gradual enrichment of nutrients (P, N) in the surface water body e.g. lake, rivers either by natural process (natural eutrophication) or anthropogenic activities (Cultural eutrophication) that result into excessive algal growth called **algal bloom,** causing several negative impact on aquatic ecosystem.Excessive growth and productivity results in decay of the biomass produced consumption of dissolved oxygen, odour production ($H_2S$), increases in pathogenic bacteria and viruses, increases in potentially toxic species of algae, fish kills, and loss of biodiversity and ultimately make water unsuitable for drinking and water supply.

Pollution of groundwater from fertilizer and animal waste mainly occurs due **nitrate leaching**. Nitrate is water soluble and negatively charged ($NO_3^-$), so it doesn't get adsorbed on soil matrix and easily leach through soils and contaminates aquifers. Nitrate contamination of water is of public concern due to two health hazard: **blue baby syndrome** or **methaenoglobinaemia** in infant of age less than 6 years and **cancer** in adult. World Health Organization (WHO) recommends that drinking water should not contain more than 1 mg $NO_3$-N per litre water (50mg, $NO_3^-$). However, nitrate is non-toxic, the concern is with its microbial reduction to nitrite. The nitrate is converted into nitrite in the intestine and then absorbed in the blood stream. Young babies cannot detoxify this nitrite, which combine with haemoglobin to form inactive form methanoglobin in thus reducing the capacity of blood to carry oxygen. Nitrate leaching is a global problem and there is ample evidence of groundwater leaching, where intense agriculture is being practised, like US, Europe, Punjab, Delhi and Andhra Pradesh.

Nearly 60,000 tonnes of pesticides are entering the Indian environment each year, of which one-third is used in public-health programmes and two-thirds in agriculture. Some of the pesticides are highly toxic and persistent, posing threat to not only human but also other non targeted organism, like birds, fishes and aquatic organisms (Table 3). The United Nations Environment Programme in 1995 identified twelve persistent organic pollutants (POPs) compounds, known as the **'dirty dozen'**, which included eight organo-chlorine pesticides (aldrin, chlordane, DDT, dieldrin, endrin, heptachlor, mirex and toxaphene), two industrial chemicals-hexachlorobenzene (HCB) and the polychlorinated biphenyl (PCB) group, and two groups of industrial by-products - dioxins and furans. These highly persistent pesticides are lipid soluble and pose risk of **bioaccumulation** and **biomagnification.** The term, bioaccumulation refers to the net accumulation over time, of pollutants within an organism from both biotic and abiotic factors. The term, biomagnifications refers to the progressive accumulation of persistent toxicants by successive trophic levels in a food chain.

Even in earlier years, the residuals of DDT, lindane and dieldrin in fish, eggs and vegetables have been much beyond the safe range in India and other countries. Long-term effects of pesticide residues in the human body include carcinogenicity, reduced life span and fertility, increased cholesterol, high infant mortality and varied metabolic and genetic disorders.

**Table 3 :** Persistence of pesticides in soil

| **Persistence** | **Half-life** | **Pesticide** |
|---|---|---|
| Non-persistent | 1-2 weeks | 2,4-D, Diquat, Endothal |
| Slightly persistent | 2-6 weeks | Dicamba, Dalapon, EPTC, Monuron, TCA |
| Moderately persistent | <6 months | Atrazine, Monuron, Linuron, Simazin, Terbacil, Trifluralin |
| Highly persistent | >6 months | DDT, HCH, Endosulfan, Aldrin |

## 3. Impact on Air Quality

Good air quality is necessary for human, plant and animal health. But, several human activities like industry, vehicle, construction, thermal power plants and agriculture are releasing variety of chemical wastes in to the atmosphere at concentration beyond its carrying capacity. This is leading to deterioration of atmosphere at not only local and regional level, but also at global scale. Air pollution has detrimental impact on human and animal respiratory system, plants metabolism and also destroy properties and monuments due to acid corrosion.

Agricultural activities like application of fertilizer and pesticide; burning of crop residue; tillage and animal husbandry have also harmful impact on air quality and human health. These activities lead to release of air pollutants like CO, $NH_3$, NOx, $SO_2$, volatile organic compounds (VOCs), particulate matter and smoke, thereby posing threat to human health. Mineralization of nitrogenous fertilizers and manure in soil produces NO, $NO_2$and $NH_3$, depending on soil Eh and pH. Crop residue burning release large amount of $CO_2$, CO, NOx, $SO_2$, volatile organic compounds (VOCs), suspended particulates matter (SPMs) while tillage operation, release soil dust particle and $CO_2$, CO, NOx, $SO_2$ and VOCs from fuel consumption in tractor. $CO_2$ and $N_2O$ are potential GHGs and responsible for global warming. $SO_2$,NO, $NO_2$ and $NH_3$ are air pollutants responsible for **acid rain**.Rain is often slightly acidic (pH = 5.6), due to the dissolution of atmospheric carbon dioxide. But, sometimes the pH of rain goes below 5.6 due to presence of acid forming pollutant gases in atmosphere like

$SO_2$, $NO_2$, and $NH_3$. These gases react with atmospheric water vapour and form strong acid like $H_2SO_4$ and $HNO_3$ and fall on the earth, in the form of acid rain or wet deposition (rain, fog, snow, hail etc.) and dry deposition (acid gas such as $SO_2$, and acidic salts such as $NH_4HSO_4$).

$$SO_2 + \frac{1}{2}O_2 + H_2O \xrightarrow[\text{ing of several steps}]{\text{Overall reaction consist-}} \{2H^+ + SO_4^{2-}\}\ (aq)$$

$$2NO_2 + \frac{1}{2}O_2 + H_2O \xrightarrow[\text{ing of several steps}]{\text{Overall reaction consist-}} 2\{H^+ + NO_3^-\}\ (aq)$$

NO and $NO_2$also react in sunlight with CO or VOCs to form **tropospheric ozone** ($O_3$) and photochemical smog [A mixture of oxidising agent like Peroxyacetyl nitrate (PAN) and ozone]. Ozone is toxic to crops, even at low concentrations, and detrimental to the health of sensitive individuals.

## 4. Impact on Ecosystem and Biodiversity

A variety of life form called biodiversity on earth makes earth ecosystem a self sustaining system. Human has greatly benefited and being benefited from ecosystem services like production of oxygen, purification of water and air, waste decomposition, soil nitrogen fixation, pollination, climate regulation etc. But, over last century greedy human is exploiting ecosystem and biodiversity beyond earth's regeneration capacity. Millennium ecosystem assessment has reported that 20% of earth's land cover has been significantly degraded; 60% of the planet's assessed ecosystems are now damaged or threatened; 35% of mangroves have been lost in the last two decades and roughly 20% of the world's coral reefs have been destroyed by several human activities. Poaching, habitat destruction spread of exotic species, pollution, and global warming are driving many species to extinction.

Despite its crucial role in feeding the world population, agriculture remains the largest driver of genetic erosion, species loss and conversion of natural habitats. Clearance of forest for agricultural land, indiscriminate use of insecticide, use of tractor and high yielding variety, monoculture and introduction of exotic species are major agricultural activities that pose threat to biodiversity. The conversion of natural habitats to cropland and other uses typically entails the replacement of systems rich in biodiversity with monocultures or systems poor in biodiversity. Large-scale agriculture brings ecosystem simplification and loss of biodiversity, thus reducing the potential to provide ecosystem services other than food production. Of some 270,000 known species of higher plants, about 10,000 – 15,000 are edible and only about 7,000 are used in agriculture. However, globalization and agricultural intensification have diminished the varieties traditionally used, with only 30% of the available crop varieties dominating global agriculture. These, together with only 14 animal species, provide an estimated 90% of the world consumed calories.

In addition to killing targeted pests, pesticides can have non-target effect on other organisms including birds, fish, beneficial insects like pollinators, predators of pests and soil microbes. It has also reported that extensive use of insecticides and herbicides has created new generations of pesticide resistant insects and plants.According to IUCN (International Union for Conservation of Nature and Natural Resources) about 4,000 of the assessed world plant and animal species are threatened by agricultural intensification. Globally, over 87% of a total of 1,226 threatened bird species are impacted by agriculture. More than 70 species are affected by agricultural pollution, 27 of them seriously. Pesticides and herbicides pose a threat to 37 threatened bird species globally, in addition to deleterious effects of agricultural chemicals on ground water.

## 5. Impact on Global Climate

Global warming and its impact on climate system i.e. climate change are emerging as biggest threat to survival of living organism on earth. It is estimated that about 15-37% of land plant and animal will be lost due to global warming by 2050. **Global warming** is increase in the average temperature of earth surface, lower atmosphere and ocean due to presence of excess amount of greenhouse gases (GHGs)like $CO_2$, $CH_4$, $N_2O$, CFCs in the atmosphere by the phenomenon called **enhanced greenhouse effect**.

GHGs are radiatively active gases, that act as transparent medium for shortwave radiation coming from sun and partial opaque medium for thermal infrared radiation (IR) emitted by earth. They allow some of IR to escape in to space (atmospheric window from 8000 to 12000 nm) and absorb majority of IR radiation and again remit it in all directions. The natural concentration of GHGs like $CO_2$, is necessary for maintaining global average earth temperature at 15°C i.e. comfortable to live and thrive otherwise in absence of GHGs the average earth surface temperature would be -18°C i.e. hostile to live and thrive. But, due to several human activities (like industry, fossil fuel consumption, widespread deforestation and burning of biomass, as well as changes in land use and land management practices) in last 100 years, the concentration of natural GHGs ($CO_2$, $CH_4$, $N_2O$ and water vapour) has increased as well as new synthetic GHGs (CFCs, HCFCs, SF6) has also entered in to the atmosphere. According to IPCC 5$^{th}$ assessment report 2014, concentrations of $CO_2$, $CH_4$, $N_2O$ has increased to 391 ppm, 1803 ppb, and 324 ppb, and exceeded the pre-industrial levels by about 40%, 150%, and 20%, respectively and global average temperature has increased by 0.85$^{o}$ Cover the period of 1880-2012.

The global warming is impacting global climate system and inducing climate change. According to IPCC, **Climate change** refers to statistically significant change in mean state of climate and/or its variability extended over long period typically a decade or more. While UN framework convention on climate change (UNFCCC) defined climate change as change in climate, which is attributed directly and indirectly to human activities that alter global atmospheric composition and which are in addition to natural climate variability observed over comparable time period. Melting of glaciers and ice cap, increasing sea level due to thermal expansion and ice melting, shorter winter and longer summer, hotter night, frequent extreme weather like flood, drought, frequent occurrence of natural calamities like cyclone are main evidence of global warming and climate change.

Out of different sources of GHGs, agriculture is one of the major contributors of GHGs in global atmosphere. According to IPCC fourth assessment report, globally about one third of the total anthropogenic increase in GHGs is attributable to agriculture and land-use changes, in which agriculture contributes about 13.5% of total global emission. According to Indian network on climate change assessment, Indian agriculture contribute 17.6% of total GHGs emission from the country and is the largest emitter of methane $CH_4$ (77.13%) and nitrous oxide $N_2O$ (60%) among all other sources. In the agriculture sub sector, $CH_4$is emitted from enteric fermentation in ruminants and rice cultivation in flooded condition; $N_2O$ emitted mainly from soil and application of synthetic fertilizer/ manure management and $CO_2$ emitted mainly due to consumption of fossil fuel during farm operation like tractor drown tillage and irrigation by motor pump.

### ***Mechanism of emission of methane and nitrous oxide from agriculture***

#### *Methane*

The global warming potential (GWP) of methane is 25 i.e. methane is about 25 times more effective than $CO_2$ as a heat-trapping gas. Wetlands, organic decay, termites, natural gas and oil extraction, biomass burning, rice cultivation, cattle and refuse landfills are the main sources and removal in the Stratosphere and soil are main sinks. Primary sources of methane from agriculture include enteric fermentation in ruminants, wetland rice cultivation and manure storage and handling.

Enteric fermentation is a microbial fermentation of feed, that takes place in the digestive systems of ruminant animals, such as cattle, sheep, and buffalo and methane is produced as a by-product of the process. Methane emission is related to the proportion of the diet (grass, legume, grain and concentrates) and the characteristics of the different feeds (e.g., soluble residue, hemicellulose and cellulose content).

Methane is also formed and emitted from submerged anaerobic condition of low land rice due to anaerobic decomposition of organic matter while upland, aerobic soil does not produce methane. The major pathways of $CH_4$ production in flooded soils, are the reduction of $CO_2$ with $H_2$ (with fatty acids or alcohols as hydrogen donor) and the transmethylation of acetic acid or methanol by highly specific methane-producing bacteria called **methanogen.**

$CO_2 + 4H_2 \rightarrow CH_4 + 2H_2O]$ Hydrogenotrophic pathway

$4CH_3OH \rightarrow 3CH_4 + CO_2 + 2H_2O$ Acetoclastic pathway

$CH_3COOH \rightarrow CH_4 + CO_2$

$4(CH_3)_3N + 6H_2O \rightarrow 9CH_4 + 3CO_2 + 4NH_3$

In paddy fields, the kinetics of the reduction processes is strongly affected by the composition and texture of soil; soil organic matter content, pH and its content of inorganic electron acceptors. The redox potential is one important factor for production of $CH_4$ in soils and it has been reported that, it must be below -150 mV in order to have $CH_4$production. It is generally recognised that, CH4 formation is only efficient in a narrow pH range around neutrality (pH from 6.4 to 7.8). Carbon substrate and nutrient availability are also important factors. Application of rice straw to paddy fields significantly increases the $CH_4$ emission rate compared with application of compost prepared with rice straw or chemical fertiliser.

Once methane is produced by methanogenes in soil, this methane is released in to atmosphere by three process namely diffusion across water surface, ebullition and through aerenchyma of rice plants. Diffusion loss of $CH_4$ across the water surface is the least important process. Methane loss as bubbles (ebullition) from paddy soils is a common and significant mechanism, especially if the soil texture is not clayey. During land preparation and initial growth of rice, ebullition is the major release mechanism. However, emission through aerenchima is main process during cropping season. Many researchers reported that more than 90 per cent of total CH4 emitted during the cropping season is released by diffusive transport through the aerenchyma system of the rice plants and not by diffusion or ebullition.

*Nitrous Oxide*

As a GHG, nitrous oxide is 298 times more effective than $CO_2$. Forests, grasslands, oceans, soils, nitrogenous fertilizers, burning of biomass and fossil fuels are the sources of nitrous oxide, while it is removed by oxidation in the Stratosphere. Soil contributes about 65% of the total nitrous oxide emission. The major sources are soil cultivation, fertilizer and manure application, and burning organic material and fossil fuels. $N_2O$ is produced in the soil as results of nitrification process under aerobic condition as well as de-nitrification process under anaerobic condition.

Organic matter (C) → $CO_2 + H_2O + NH_4$ (Ammonification)

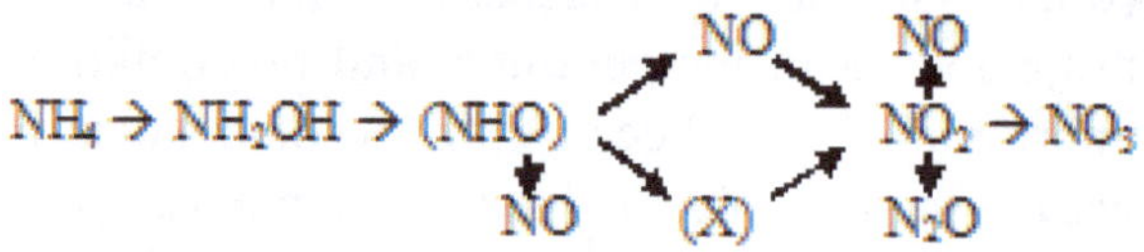

(Nitrification, aerobic)

$NO_3 \rightarrow NO_2 \rightarrow NO \rightarrow N_2O \rightarrow N_2$ (Denitrification, anaerobic)

## Impact of Climate Change on Agriculture

Agriculture production is highly dependent on state of environment. The state of environment has effects on food production through its role in water, nutrients, soils, climate,and weather,abundance of pathogens, weeds and pests as well as on insects, that are important for pollination. Environmental degradation due to unsustainable human practices and activities, now seriously endangers the entire production platform of the planet. According to a report of United Nation Environment Programme (UNEP) up to 25% of the world's food production may be lost due to environmental breakdown by 2050 unless action is taken. Climate change is emerging as the biggest threat for global agriculture and other sector. It may have direct or indirect impact on food production (Table 4). The increase in atmospheric carbon dioxide has a direct fertilization effect on crops with $C_3$ photosynthetic pathway, and thus promotes their growth and productivity. Increase in temperature can reduce crop duration, increase crop respiration rates, alter photosynthate partitioning to economic products, effect the survival and distributions of pest populations, hasten nutrient mineralization in soils, decrease fertilizer use efficiencies, and increase evapo-transpiration. Indirectly, there may be considerable effects on land use due to snow melt, availability of irrigation, frequency and intensity of inter- and intra-seasonal droughts and floods, soil organic matter transformations, soil erosion, changes in pest profiles, decline in arable areas due to submergence of coastal lands, and availability of energy. Equally important determinants of food supply are socio-economic environment including government policies, capital availability, prices and returns, infrastructure, land reforms, and inter- and intra-national trade, that might be affected by climatic change.

According to IPPCC fourth assessment report, potential for global food production may rise with increases in local average temperature over a range of 1–3°C (before 2050), but above this range (after 2050) may decrease. Model projections suggest that although increased temperature and decreased soil moisture will act to reduce global crop yields by 2050, the direct fertilization effect of rising carbon dioxide concentration (CO2) will offset these losses.

Furthermore, projected changes in the frequency and severity of extreme climate events are predicted to have more serious consequences for food and food security than changes in projected mean temperatures and precipitation. Developing countries are more vulnerable, because of the dominance of agriculture in their economies, the scarcity of capital for adaptation measures, their warmer baseline climates and heightened exposure to extreme events.

**Table 4** : Probable impacts of climate change on various sectors of Indian agriculture

| **Sector** | **Impact** |
|---|---|
| Crop | • Yields of major cereals crops, especially wheat to be reduced due to increase in minimum temperature during the reproductive period. |
| | • Increase in extreme weather events adversely affecting agricultural productivity. Reduction in yields in the rainfed areas due to changes in rainfall pattern during monsoon season and increased crop water demand. |
| | • Yield loss of thermo-sensitive crops such as basmati rice, cabbage, cauliflower, apple, peach, cherry and saffron. |
| | • Quality of fruits, vegetables, tea, coffee, aromatic, and medicinal plants to be affected. |
| | • Pest and diseases of crops to be altered because of more enhanced pathogen and vector development, rapid pathogen transmission and increased host susceptibility. |
| | • Agricultural biodiversity is also threatened by decreased rainfall and increased temperature, sea level rise, and increased frequency and severity of drought, cyclones and floods. |
| Water | • Demand for irrigation to increase with increased temperature and higher amount of evapo-transpiration. This may also result in lowering of groundwater table at some places. |
| | • The melting of glaciers in the Himalayas will increase water availability in the Ganges, Bhramaputra and their tributaries in the short run but in the long run the availability of water will decrease considerably. |
| | • The water balance in different parts of India will be disturbed and the quality of groundwater along the coastal track will be more affected due to intrusion of sea waters. |
| Soil | • Organic matter content, which is already quite low in Indian soil, would become still lower. Quality of soil organic matter may be affected. |
| | • The residues of crops under elevated $CO_2$ concentration will have higher C:N ratio, and this may reduce their rate of decomposition and nutrient supply. |
| | • Increase of soil temperature will increase N mineralization, but its availability may decrease due to increased gaseous losses through processes, such as volatilization and denitrification. |
| | • Change in rainfall volume and frequency, and wind may alter the severity, frequency and extent of soil erosion. |

| | |
|---|---|
| | • Increased temperature coupled with reduced rainfall may lead to upward water movement resulting in accumulation of salts in upper soil layers.<br>• Similarly, rise in sea level may lead to salt-water ingression in the coastal lands turning them unsuitable for conventional agriculture.<br>• Biological N fixation under elevated concentrations of $CO_2$ may increase, provided other nutrients are not limiting. |
| Livestock | • Climate change will affect feed production and nutrition of livestock. Increased temperature would increase lignification of plant tissues reducing the digestibility. Increased water scarcity would also decrease food and fodder production.<br>• Major impacts on vector-borne diseases through expansion of vector populations into cooler areas. Changes in rainfall pattern may also influence expansion of vectors during wetter years, leading to large outbreaks of disease.<br>• Increases in heat-related mortality and morbidity. |
| Fishery | • Temperature changes are likely to impact cool water species negatively, warm water species positively.<br>• Impacts of increased temperature and tropical cyclonic activity would affect the capture, production and marketing costs of the marine fish.<br>• Migration of different marine and inland species to favourable climate regions. |

## Environmental Legislation, Standards and Monitoring

Gradually, people are realising the importance of environment and raising concern about it. At global platform, many initiatives have been taken and time to time global convention and protocol (like Montreal protocol for eliminating the use and production of ozone depleting substances; Kyoto protocol for curbing the emission of GHGs) has been formulated and applied for environmental protection. Indian constitution also provides legislative measures for the environmental protection and many acts and rules have been formulated and implemented.

### *1. Environmental legislation in India*

As per the Constitution of India (Art-48), it is the duty of the state 'to protect and improve the environment and to safeguard the forests and wildlife of the country'. Article 51A(g), imposes a duty on every citizen 'to protect and improve the natural environment including forests, lakes, rivers, and wildlife'. Reference to the environment has also been made in the Directive Principles of State Policy as well as the Fundamental Rights. The Department of Environment was established in India in 1980, to ensure a healthy environment for the country. This became the Ministry of Environment and Forests in 1985. The constitutional provisions are backed by a number of Laws – Acts, Rules, and Notifications. The EPA (Environment Protection Act), 1986 came into force soon after the Bhopal Gas Tragedy and is considered 'umbrella legislation' as it filled many gaps in the existing laws. Thereafter, a large number of laws came into existence as the problems began rising.

- **The Water (Prevention and Control of Pollution) Act(1974)** establishes an institutional structure for preventing and abating water pollution. It establishes standards for water quality and effluent. As per the Act, the polluting industries must seek permission to discharge waste into effluent bodies. The CPCB (Central Pollution Control Board) and the State Pollution Control Boards were constituted for implementing this act.
- **The Air (Prevention and Control of Pollution) Act (1981)** provides guidelines for the control and abatement of air pollution. It entrusts the power of enforcing this act to the CPCB.
- **The Environment (Protection) Act (1986)** authorizes the central government to protect and improve environmental quality, control and reduce pollution from all sources, and prohibit or restrict the setting and/or operation of any industrial facility on environmental grounds.
- **The Biological Diversity Act (2002)** is an act to provide for the conservation of biological diversity, sustainable use of its components, and fair and equitable sharing of the benefits arising out of the use of biological resources and knowledge associated with it.

***2. Indian environmental standards***

An environmental standard is a policy guideline, that regulates the effect of human activity upon the environment. They are a set of quality conditions (like pH, EC, emission of SPM etc.), that are adhered or maintained for a particular environmental component and function. Environmental standards not only protect people and the environment, they also secure a consistent approach and decision-making process throughout the country. The different environmental activities have different concerns and therefore have different standards. In India, central pollution control board (CPCB) under ministry of environment is nodal agency for publishing guideline and standard for environmental protection.CPCB has developed many national standards like ambient air quality, water use criteria, effluents and emission standards for different industries, standards for vehicular exhaust etc.These standards have been approved and notified by the Government of India, Ministry of Environment & Forests, under Section 25 of the Environmental (Protection) Act, 1986. The water and air quality standards of CPCB and standard methods of estimation of water and air quality parameters are given in (Table 5-7).

**Table 5** : Quality parameters of fresh water for different use

| Use | Class | Parameter | Value |
|---|---|---|---|
| Drinking water without conventional treatment but after disinfections | A | Total coliform organisms per 100 ml | <50 |
| | | pH | 6.5-8.5 |
| | | Dissolved oxygen (mg $l^{-1}$) | >6 |
| | | Biochemical oxygen demand (mg $l^{-1}$) | <2 |
| | | Total dissolved solids (mg $l^{-1}$) | <500 |
| | | Chlorides (Cl) (mg $l^{-1}$) | <250 |
| | | Colour (Hazen unit) | <10 |
| | | Sulphates ($SO_4$) (mg $l^{-1}$) | 400 |
| | | Nitrate ($NO_3$) (mg $l^{-1}$) | 45 |
| | | Arsenic (As) (mg $l^{-1}$) | 0.05 |
| | | Iron (Fe) (mg $l^{-1}$) | 0.3 |
| | | Fluorides (F) (mg $l^{-1}$) | 1.5 |
| | | Lead (Pb) (mg $l^{-1}$) | 0.1 |
| | | Copper (Cu) (mg $l^{-1}$) | 1.5 |
| | | Zinc (Zn) (mg $l^{-1}$) | 15 |
| Outdoor bathing (organised) | B | Total coliform organisms per 100 ml | <500 |
| | | pH | 6.5-8.5 |
| | | Dissolved oxygen (mg $l^{-1}$) | >5 |
| | | Biochemical oxygen demand (mg $l^{-1}$) | <3 |
| Drinking water with conventional treatment | C | Total coliform organisms per 100 ml | <5000 |
| | | pH | 6.0-9.0 |
| | | Dissolved oxygen (mg $l^{-1}$) | >4 |
| | | Biochemical oxygen demand (mg $l^{-1}$) | <3 |
| | | Nitrate ($NO_3$) (mg $l^{-1}$) | 20 |
| Propagation of wild life. fisheries | D | pH | 6.5-8.5 |
| | | Dissolved oxygen (mg $l^{-1}$) | >4 |
| | | Free ammonia (as N) (mg $l^{-1}$) | 1.2 |
| Irrigation, industrial cooling, controlled waste disposal | E | pH | 6.5-8.5 |
| | | Electrical conductivity (mmhos $cm^{-1}$) | <2.25 |
| | | Total dissolved solids (mg $l^{-1}$) | <2100 |
| | | Chlorides (Cl) (mg $l^{-1}$) | <600 |
| | | Sodium absorption ratio (SAR) | <26 |
| | | Boron (mg $l^{-1}$) | 2 |
| | | Sulphates ($SO_4$) (mg $l^{-1}$) | 1000 |

**Table 6** : Methods for estimation of water quality parameters

| Water parameters | Methods of estimation |
|---|---|
| Microbes | • Multiple tube fermentation technique for coliform bacteria (MPN test) |
| | • Membrane filtration method for total coliform and thermotolerant 31 (faecal) |
| | • EC-MUG Test for confirmation of E. coli |
| Boron (B) | • Curcumin Method |
| | • Carmine Method |
| Phenols | • Chloroform Extraction Method |
| | • Direct Photometric Method |
| pH and Electrical conductivity | • PH and EC meter |
| Dissolved oxygen | • The Winkler Method with azide Modification· Membrane Electrode Method |
| Biological Oxygen Demand (BOD) | • Titrimetric Method |
| | • Respirometric Method |
| Chemical oxygen demand (COD) | • Open Reflux Method |
| | • Closed Reflux (titrimetric and colorimetric) Method using COD Digester |
| Phosphate | • Stannous Chloride Method |
| Ammonia | • Nesslerisation Method |
| Nitrate | • UV Spectrophotometric Method |
| | • Phenol Disulphonic Acid (PDA) Method |
| Nitrite | • Colorimetric Method |
| Sulphate | • Turbidity method |
| Fluoride | • Ion Selective Electrode Method |
| | • Sodium 2-(parasulphophenylazo)-1,8-dihydroxy-3,6 Naphthalene 128 Disulphonate (SPADNS) Method |
| Chloride | • Argentometric method |
| Cyanide | • Colorimetric |
| Oil and Grease | • Partition-gravimetric method |
| | • Soxhlet extraction method |
| Metals | • Inductively coupled Plasma Method |

**Table 7 :** National ambient air quality standard

| **Pollutant** | **Time weighted average** | **Concentration in ambient air** | | **Methods of measurement** |
|---|---|---|---|---|
| | | **Industrial, Residential, Rural and other area** | **Ecologically sensitive area (notified by Central Gvt.)** | |
| $SO_2$ (µg/m$^3$) | Annual | 50 | 20 | A. Improved West and Gaeke |
| | *24 hours** | 80 | 80 | B. Ultraviolet fluorescence |
| $NO_2$ (µg/m$^3$) | Annual | 40 | 30 | A. Modified Jacob and Hochheiser |
| | *24 hours** | 80 | 80 | (Na Arsenate) |
| | | | | B. Chemiluminescence |
| $PM_{10}$ (µg/m$^3$) | Annual | 60 | 60 | A. Gravimetric |
| | *24 hours** | 100 | 100 | B. TOAM |
| | | | | C. β Attenuation |
| $PM_{2.5}$ (µg/m$^3$) | Annual*24 hours** | 4060 | 4060 | A. Gravimetric |
| | | | | B. TOAM |
| | | | | C. β Attenuation |
| Lead (Pb) | Annual | 0.5 | 0.5 | A. AAS/ICP (sampling |
| (µg/m$^3$) | *24 hours** | 1 | 1 | EPM 2000 or equivalent filter paper) |
| | | | | B. ED-XRF using Teflon filter |
| $NH_3$ (µg/m$^3$) | Annual | 100 | 100 | A. Chemiluminecence |
| | *24 hours** | 400 | 400 | B. Indophenols blue method |
| Ozone ($O_3$) | 8 hours | 100 | 100 | A. UV Photometric |
| (µg/m$^3$) | *1 hours** | 180 | 180 | B. Chemiluminescence |
| | | | | C. Chemical method |
| CO (µg/m$^3$) | 8 hours<br>*1 hours** | 2<br>4 | 2<br>4 | A. Non dispersive infrared (NDIR) spectroscopy |
| Benzene | Annual | 5 | 5 | A. Gas chromatography based continuous analyzer |

*Contd...*

| | | | | |
|---|---|---|---|---|
| | | | | B. Adsorption and desorption followed by GC |
| Benzo(o) Pyrene PM | Annual | 0.001 | 0.001 | A. Solvent extraction followed by HPLC/GC |
| Arsenic (As) (μg/m$^3$) | Annual | 0.006 | 0.006 | A. AAS/ICP (sampling EPM 2000 or equivalent filter paper) |
| Nickel (Ni) (μg/m$^3$) | Annual | 0.02 | 0.02 | B. AAS/ICP (sampling EPM 2000 or equivalent filter paper) |

* Annual arithmetic mean of minimum 104 measurements in a year at a particular site taken twice a week 24 at uniform intervals

** should be complied with 98% of the time in a year, 2% of time, they may exceed the limit but not on two consecutive day of monitoring.

## 3. Environmental Monitoring

Environmental monitoring is systemic and scientific observation of changes, that occur in environment over certain period of time and space. It gives idea about the current status of environment, level of pollutants, trends of change in environment and also provides early warning for any risk arising due to sudden rise of pollutant level like air pollution. In order to ascertain the potential or actual extent of pollution that has occurred, it is necessary to undertake environmental monitoring of the polluted site.Following points should be kept in mind during developing a program to monitor the extent or effects of pollution in a particular environment

- Purpose of monitoring : this includes determining, for example, pollution level changes in space and time
- Objectives: which may include specific chemical, physical, or biological analyses and,
- Approach: which will assist in defining the number and types of measurements
- Knowledge about environmental characteristics or uniqueness of each environment
- Sampling methods, including locations, timing, and type
- Knowledge about basic concepts and applications of analysis and,

measurement principles.

- All data must be reported with appropriate degree of precision by considering the sampling methods, instrumental methods, and number of samples analysed.
- Environmental data should be reported using commonly accepted units, preferably conforming to the SI (International System of Units), used worldwide.
- Statistics and geo-statistics: Statistical methods are necessary in pollution science, because it is impossible to characterize all properties of an environment, everywhere all the time. Statistics are used to select samples from a population in an unbiased manner. They also help to interpret the data with the appropriate degree of confidence.

***Environmental monitoring tools***

There several modern and advanced monitoring tools, that are being employed to get more data and clear cut idea about the state of environment. Some of the basic tools are different types of maps, aerial photographs, remote sensing, sampling tools and automatic monitoring systems. **Maps** are small portable representations of a portion of our environment, which helps us in locating a place interest and give idea about the land features. All maps have a scale, which is the ratio between the map's distances and the real (earth) distances. Thus, most scales are a unit less ratio or fraction of two distances. For example, if 1 cm on the map represents 1 km in reality, then the map scale is 1/100,000. There are three major types of maps: **Planimetric maps**, which present scaled information in two dimensions; **Topographic maps**, which add elevation information to land maps with the use of contour lines that is defined as equal elevations drawn at fixed elevation intervals.Contour lines are very useful in identifying relief and unique land features such as watersheds, rivers, depressions, and mountains. **Thematic maps** or geographic information system (GIS) maps are very useful in environmental applications, because they provide a way to arrange and present layers of data or themes as maps.Thus, GIS maps are decision-making tools, that provide a visual representation of large amounts and types of data, which can be easily sorted because all the data and coordinates are in a digital form. Because GIS uses digital data that can be manipulated using Boolean operations (binary 0 and 1 values), data can be re-classified, sorted, overlaid, and used to measure distances.

**Remote Sensing** is an advanced environmental monitoring tool. It is a method of obtaining information about an object from remote location without actual physical contact with object e.g. aerial photography or satellite mounted sensors. Presently, space-based sensors, usually satellite mounted, are being used to observe biophysical and geochemical phenomena in earth's atmosphere and land/ocean surfaces.Most remote sensors measure light from different parts of the electromagnetic spectrum that is reflected from earth's atmosphere and surface. The UV region is commonly used to measure air pollutants. Remote sensing imagery is captured using digital cameras that have finite optical pixels and therefore finite spatial resolution in terms of surface area per pixel. Sensors are rated in terms of coarse, medium or fine resolution. For example, high-resolution sensors are able to collect data ranging 100–1 m in resolution. These sensors are commonly used to study land cover patterns and changes that result from human activities, including agriculture, plant disease, urbanization, and deforestation. Coarse resolution sensors provide data ranging from 100 to1000 m in resolution and are useful to study cloud cover, dust, and sediment transport. Other remote-sensing applications include observing and measuring vegetation, snow cover changes, weather changes and forecasting, and physical land degradation, including erosion.

***Environmental sampling tools***

Sampling is the ut most important component in environmental monitoring. The objective of sampling is to collect representative sample i.e. sample in which relative proportions or concentration of all pertinent components will be the same as in the material being sampled. Faulty sampling procedure may lead to collection of sample, which may misguide the result from true status of environmental condition. So, sampling should be properly planned and conducted. Before sampling flowing point, should be considered:

- **Sampling strategies**: Number and type of samples, locations, depths, times, intervals.
- **Sampling methods**: Specific techniques and equipment to be used.
- **Sample storage:** Types of containers, preservation methods, maximum holding times.
- **Sample recording:**identification number, location, sample collector, date and time, sample type (Grab or composite). The information may be attached tag, labelling or writing on container with water proof ink.

**Safety Considerations**

- Always prohibit eating, drinking, or smoking near samples, sampling locations, and in the laboratory.
- Keep sparks, flames, and excessive heat sources away from

samples and sampling locations.

- If flammable compounds are suspected or known to be present and samples are to be refrigerated, use only specially designed explosion-proof refrigerators.
- Label adequately any sample known or suspected to be hazardous, because of flammability, corrosivity, toxicity, oxidizing chemicals, or radioactivity, so that appropriate precautions can be taken during sample handling, storage, and disposal.

### *Soil and the subsurface soil sampling*

Physically, surface soil samples are easy to obtain using inexpensive equipment such as a shovel or a soil auger. Augers are useful in that they allow samples to be taken at exactly the same depth on every occasion. Augers and sampling tubes are commonly used to collect soil root zone (1.5 m depth) samples. Sampling tubes are more appropriate to collect soil samples for pollution-related measurements as they include an inert plastic or metal liner, that is used to protect and store the sample.

Subsurface sampling is more complex and expensive than surface soil sampling, particularly when deep subsurface sampling is attempted. For subsurface sampling, mechanical drill rigs, such as rotary mud drilling or hollow-stem augers are necessary. The hollow stem auger is used to drill to the sampling depth by engine power. Then, a center rod that plugs the hollow stem during drilling is removed and a clean sampling tube is inserted to collect a soil sample.

Because soils are heterogeneous, it is better to collect multiple cores that are mixed together to give a composite sample. Polluted soils should be sampled in an unbiased manner (random sampling) if the source of the pollution is not known. Systematic and directional soil sampling may be more appropriate in cases, where the source of pollution is known and may be correlated to spatial location. Soil samples that collected for microbial analysis should be kept on ice while transported to the laboratory. Microbial analyses should be performed as soon as possible to minimize the effects of storage on microbial populations and should not be air dried prior to analyses. Soils sampled for chemical analysis should be air dried and can then be kept indefinitely pending analysis.

### *Soil pore water sampling*

Soil pore water is often sampled to monitor movement of water-soluble pollutants in the sub surface zone. Generally, suction lysimeter and porous cup is used for this purpose. The use of suction lysimeters extends to industrial municipal and agricultural waste treatment and disposal sites (landfills, ponds, septic systems, wetlands, irrigation). Porous cups made of ceramic or stainless steel is inserted into the soil by drilling a hole. The types of pollutants that can be monitored using these devices include salts, nitrates, soluble metals (like CrVI), and soluble organic carbon compounds.

### *Surface water sampling*

Monitoring pollutants in water has become a global concern, due to the limited and diminishing water resources and the negative impact of pollutants on ocean and fresh water sources. Collecting water samples tends to be somewhat easier than sampling soils. First of all, water at a given site tends to be more homogenous than soils, with less site-to-site variability between two samples collected within the same vicinity. Secondly, it is often physically easier to collect water samples. Field portable devices are commonly used to measure (*in situ*) major water quality parameters such as pH, TDS, oxygen saturation, temperature, and turbidity. Water sampling followed by analysis in the laboratory is common to monitor specific pollutants such as trace organic and inorganic chemicals (pesticides, hydrocarbons, toxic metals), and pathogens like E. colibacteria. Surface water samples can be collected in wide-mouth polyethylene jars or with a bucket. The amount of water collected can be a few milliliters, such as when routine analyses such as pH are to be done. In other cases, large volumes need to be collected (1000 liters), as in the case of determining the presence of enteric viruses in marine waters. Normally, water samples are kept as cool as possible in sealed containers to prevent microbial and chemical activity and preclude evaporation.

### *Air sampling*

The atmosphere is a particularly difficult analytical system, because of the very low levels of substances to be analysed; sharp variations in pollutant level with time and location; differences in temperature and humidity; and difficulties encountered in reaching desired sampling points, particularly those substantially above the earth's surface. However, the collection of air samples for analysis can be done in a variety of ways. In some cases, samples of air are diverted automatically into instruments for continuous measurement of pollutant concentrations. In yet other applications, air samples are collected in sample bags for later laboratory chemical analysis.

A number of factors enter into obtaining a good air sample. The size of the sample required (total volume of air sampled) decreases with increasing concentration of pollutant and increasing sensitivity of the analytical method. Often, a sample of ten or more cubic meters is required. The sampling rate is determined by the equipment used and generally ranges from approximately

0.003 $m^3$/min to 3.0 $m^3$/min. Samples can be divided into two major categories. **Grab samples** are taken at a single time and in a single place. Therefore, they are very specific with respect to time and location. **Composite samples** are collected over an extended time and may encompass different locations as well. In principle, the average results from a large number of grab samples give the same information as a composite sample. A composite sample has the advantage of providing an overall picture from only one analysis. On the other hand, it may miss extreme concentrations and important variations that occur over time and space.

The most straight forward technique for the collection of particles is sedimentation. A **sedimentation** collector may be as simple as a glass jar equipped with a funnel. Liquid is sometimes added to the collector to prevent the solids from being blown out. **Filtration** is the most common technique for sampling particulate matter. For this purpose, high volume gas sampler is used. Filters composed of fritted (porous) glass, porous ceramics, paper fibers, cellulose fibers, fiber glass, asbestos, mineral wool, or plastic may be used. A special type of filter is the membrane filter, which yields high flow rates with small, moderately uniform pores. Other device used to collect aerosols, include impingement and impaction devices. **Impingement** is the trapping of airborne particles in a liquid matrix. In contrast, **impaction** is the forced deposition of airborne particles on a solid surface.

Sampling for vapors and gases may range from methods designed to collect only one specific pollutant, to those designed to collect all pollutants. Essentially all pollutants may be removed from an air sample **cryogenically** by freezing or by **liquefying** the air in collectors maintained at a low temperature. **Absorption** in a solvent, such as by bubbling the gas through a absorbing liquid can be used for the collection of gaseous pollutants. **Adsorption**, in which a gas collects on the surface of a solid, is particularly useful for the collection of samples to be analysed by gas chromatography.

## 4. Greenhouse Gas Sampling and Field Measurement

Methane, nitrous oxide and carbon dioxide are major GHGs coming out from agricultural activities. These gases can be manually or automatically sampled followed by analysis. Some of methods include closed chamber method, open chamber method and micro-meteorological method. Closed chamber technique is commonly used for sampling of GHGs from agricultural field.Chamber of suitable dimension (generally 50 cm x 30 cm x 100 cm) can be made from inert material like rigid plastic, perspex or acrylic sheet. Gas flux from the soil can be determined by collecting gas samples, periodically from the chamber and measuring the change in concentration of a gas with time during the period of linear concentration change (generally from zero hour to two hours at fixed interval after air tight fixing of chamber in soil). The gas sample can be drawn by using 20-50 ml syringe fitted with 24 gauge hypodermic needle and three way stop cock. Concentration of sampled gas is immediately determined by using gas chromatography (GC) fitted with different detectors. Nitrous oxide is analysed by electron capture detector (ECD), while methane is analysed by flame ionization detector (FID). Carbon dioxide can be directly quantified in field itself by IRGA (infra red gas analyser) or can be sampled and analysed with GC fitted with methanizer followed by FID detector.

Eddy covariance technique a microclimatic method of GHGs flux measurement is getting popular due to its potential for very short response time, continuous *in situ* measurement over large area and generation of reliable data. This technique is being used to measure vertical turbulence flux of water, carbon dioxide, methane, heat, ozone and volatile organic compound, using different analyser and sensors in the atmospheric boundary layer.

## 5. Assessment of Impact of Climate Change on Crops

There are several approaches that can be used to study the impact of climate change on crops. Most common approaches are long term surveys, experiments and simulation modelling.

### *Long term survey*

It can be used to collect data related to shift in occurrence of phenological events in crops, shift in cropping pattern, and shift in distribution ranges of insect species etc. by regular survey at discrete locations characterized by different latitudes and altitudes. The collected data can be used to correlate and analyse the impact of rising temperature, declining water availability etc. on crop.

### *Experiments*

The experiment may be conducted to see the direct impact of climate change on crop by creating different climatic condition. To understand the impacts of elevated levels of $CO_2$, $O_3$ and temperature, a number of methodologies have been developed since the 1970s. To simulate the effects of elevated $CO_2$ concentrations on plants, initially methods used are leaf cuvettes, plant grow chambers and greenhouses. However, these methods have many constraints in plot and plant size, and require active control of all environmental variables. To overcome some of these limitations, other advanced methodologies, which are widely accepted for such studies are:

(a) Open Top Chamber (OTC),

(b) Free Air $CO_2$ Enrichment (FACE)

(c) Temperature gradient tunnels (TGT)

**Open Top Chambers:** OTCs are made of a plastic enclosure (of Poly vinyl chloride sheets) with inclined walls at top. Top of the structure is kept open to allow the natural air exchange and maintain normal temperature and humidity. Air enriched with particular gas of interest ($CO_2$ and $O_3$)from a given source enters near the bottom and distributed inside with the help of circular tubes fitted at the base of OTC and flows out the open top creating an enriched $CO_2$and $O_3$ environment inside the chamber. Desired level of concentration of gas in interest, are automatically controlled and maintained by sensors and computer programme.

### *Free Air $CO_2$Enrichment (FACE)*

OTC alters microclimatic condition surrounding crops due to presence of side wall. This draw back led to development of wall less enrichment system called FACE. FACE has minimal impact on micro climate of crops, however it requires more amount of carbon dioxide for enrichment due to wastage in absence of side wall. A typical FACE system comprises of an approximately circular facility and is surrounded by an array of vertical or horizontal pipes that release air enriched with $CO_2$ at vertical intervals from just above the ground to vicinity of the plant canopy. The FACE ring (plenum) is made up of eight horizontal PVC pipes fitted with solenoid valve. Opening and closing of the solenoid valves in the rings are regulated by the speed and direction of wind. $CO_2$is released from the pipes present in the upwind direction at any given time. Desired level of concentration of gas in interest are automatically controlled and maintained by sensors and computer programme.

### *Temperature Gradient Tunnel (TGT)*

To study the impact of rising temperature on crop, temperature gradient tunnels are a good option, because it provides a range of temperature inside. TGT consists of a series of semicircular bows anchored in the soil. The bows are covered with polycarbonate sheet or UV stabilized polythene sheet to create a greenhouse like atmosphere, having a temperature higher than the outside environment. Air inside the TGT is heated naturally by the incident solar radiation during the daytime and thermal gradient is created by sucking ambient air into the tunnel from one end (inlet) and releasing it from the other end (outlet) by fan. Desired level of temperature gradient is achieved by the regulation of fan speed. Temperature and humidity sensors are installed inside the tunnel at regular intervals, to measure air temperature and humidity at different points of the tunnel and data are recorded by data logger connected with computer. Operational for 24 hours (day and night) throughout the crop growing period and temperature gradient data are recorded during this period.

### *Crop growth simulation models*

Model is simplified representation of a system or a process. Simulation modelling is the process of developing a model based on mathematical equation that mimics the process and function of natural system. It helps in understanding the complex nature of natural system and interaction among them, which is not possible by a single or few experiments. Once model is developed, calibrated and validated, it helps in generating a lot of information about present or future scenario in short period of time without going for long term experiment. Impact of climate change on crop will not depend on single factor but would depend upon on complex interactions of climatic and biological factors with technological and socioeconomic changes that are difficult to predict. Hence, quantitative modeling approaches, which allow investigating multiple scenarios and interactions simultaneously is becoming more important for impact assessment.

A crop model is a simple representation of crop and its interaction with environment. These are based on quantitative understanding of the underlying crop physiological, meteorological and soil physical, chemical or biological processes and integrate the effect of soil, weather, crop, pests and management factors on growth and yield of crops. Several crop growth simulation models viz., CERES-Rice; CROPGRO models for grain legumes; ORYZA for rice; SUCROS, a generic model parameterized for wheat, potato, maize; WTGROS for wheat; InfoCrop, a generic crop growth model parameterized for rice, wheat and other crops were developed with a purpose of looking in to opportunities of improving crop production.

## Conclusion

Good environmental condition is necessary for the sustainability of life on earth. There is ample evidence that our environment has been degraded in past and continue to be degraded in future by several human activities. Agriculture is one of the major causes for environmental degradation and is also highly susceptible and vulnerable to environmental change. The emerging global threat like climate change, ozone depletion, loss of biodiversity, air pollution will have several harmful impact on agricultural production system, that will ultimately affect socio-economic condition of the global society. But, it's not too late; our present activity will decide the fate of future. There is need for wise planning and efficient utilization of natural resources for sustainable development, bright and healthy future ahead.

## References


Aktar, M.W., Sengupta, D. and Chowdhury, A. 2009: Impact of pesticides use in agriculture: their benefits and hazards. *Interdisciplinary Toxicology*. 2(1): 1–12.

Atafar, Z., Mesdaghinia, A., Nouri, J., Homaee, M., Yunesian, M., Ahmadimoghaddam, M., Mahvi, A.H. 2010. Effect of fertilizer application on soil heavy metal concentration. *Environment Monitoring and Assessment*. 160:83–89.

Ellis, E.C., Kaplan, J.O., Fuller, D.Q., Vavrus, S., Goldewijk, K.K. and Verburg, P.H. 2013. Used planet: A global history.*PANAS*. 110 (20): 7978–7985. www.pnas.org/cgi/doi/10.1073/pnas.1217241110

INCCA, 2011. India: Green House Gas Emission 2007: Indian network for climate change assessement, Ministry of Environment and Forests. Government of India.

IPCC, 2013 : Summary for Policy makers. *In*: Climate Change 2013: The Physical Science Basis. Contribution of Working Group I, II and III to the Fifth Assessment Report of the Inter Governmental Panel on Climate Change.

IPCC, 2007. Summary for Policymakers. In: Climate Change 2007: The Physical Science Basis. Contribution of Working Group I, II and III to the Fourth Assessment Report of the Intergovernmental Panel on Climate Change.

Jain, N., Bhatia, A. and Pathak, H. 2014. Emission of Air Pollutants from Crop Residue Burning in India. *Aerosol and Air Quality Research*. 14: 422–430.

Lal, R., 1993. Tillage effects on soil degradation, soil resilience, soil quality, and sustainability. *Soil and Tillage Research*. 27: 1–8.

MEA, 2005. Millennium Ecosystem Assessment. Ecosystems and Human Well-being: Synthesis. – Island Press, Washington, DC.

SALOW, 2011. Summary report on "The State of the World's Land and Water Resources for Food and Agriculture", Food and Agricultural Organization, Roam.

Savci, S. 2012. An Agricultural Pollutant: Chemical Fertilizer. *International Journal of Environmental Science and Development*. 3 (1): 77-80.

Singh, R.B. 2000. Environmental consequences of agricultural development: a case study from the Green Revolution state of Haryana, India. Agriculture, *Ecosystems and Environment*. 82: 97–103.

State of Environment Report 2009. Environmental Information System (ENVIS), Ministry of Environment & Forests, Government of India.


❑❑❑

CHAPTER - 8

# Soil Nutrient Mapping - A Case Study

**A.K. Sahoo, S.K, Singh, Dipak Sarkar and A.K. Sarkar**

Spatial variability in soil parameters including nutrients has been attributed to the parent material, topography, landforms, cropping pattern, fertilization and manuring practices.. Blanket nutrient recommendations further widen the variability and enhance the risk of soil degradation in terms of soil organic carbon and nutrient mining, acidity, green house gas emission, water and environmental pollution. Unfavorable economics on account of blanket recommendation, adversely influence farmers to increase investments in new crop management technologies. This results in low farm productivity and poor soil health, that jeopardize food security and agricultural sustainability.

Site Specific Nutrient Management (SSNM) is one of the logical steps to overcome the problems, but inadequate infrastructure for soil testing facilities in the country limits its practical implementation. Soil parameter mapping including nutrients involving Geo-referenced soil sampling, laboratory analysis, database structuring in GIS and subsequent interpolation using appropriate geo-statistics provides a valid ground for SSNM. The possibility of using such maps as a decision support tool, to guide nutrient application in a site specific mode need to be explored. Nutrient sufficiency/deficiency zones of single and multiple nutrients prepared using logical algorithms in GIS are one step forward for SSNM (NBSS&LUP 2006, Sen *et al* 2012, Sahoo *et al* 2013). Repetitive survey using these maps can give a clear visual indication of the changing fertility scenario at block/village/district level on real time scale. This kind of information is imperative for the area affected with wide spread acidification, single and multiple nutrient deficiency, water stress and intermittent drought spells during cropping season.

## Methodology

### *Preparation of base map*

A reliable base map of the study area is to be prepared for collection of soil samples from a particular site / point. In general, the following base maps are used for collection of soil samples at different levels of soil nutrients mapping:

- Village/ Farm/ Micro-watershed - Cadastral maps (1:3960 scale) or available map at desired scale of the study area.
- Block / PS. / Sub-watershed - Available map at desired scale of the Block / PS/ Sub-watershed
- District / Watershed – Survey of India map ( 1: 50,000 scale ) for the area of respective district / watershed.

The final base map of the study area is to be prepared with demarcation of grid points (soil sampling point) at specific interval (as desired).

***GPS Based soil sampling***

The soil sample must perfectly represent the area. Variation in slope, colour, texture, crop growth and management should be taken into account and separate sets of composite samples need to be collected from each of such area. A field can be treated as a single sampling unit ,only if it is more or less uniform in all respect.

From soil fertility point of view, normally the composite samples from each grid point is taken from the plough layer i.e. up to 0-15 cm/ 0 – 20 cm/ 0-25cm depth using **GPS** to identify actual locations on the map. This is applicable for the fields growing cereals and other seasonal crops. In case of deep rooted crops like sugarcane and dry farming conditions, it may be necessary to obtain samples from different depths or layers of soil. Geographical location of each sampling point is collected through Global positioning system (GPS).

As an example, total five samples (Fig.1) are mixed thoroughly to have one composite sample, of which a pack of one kilogram sample is retained for laboratory investigations.

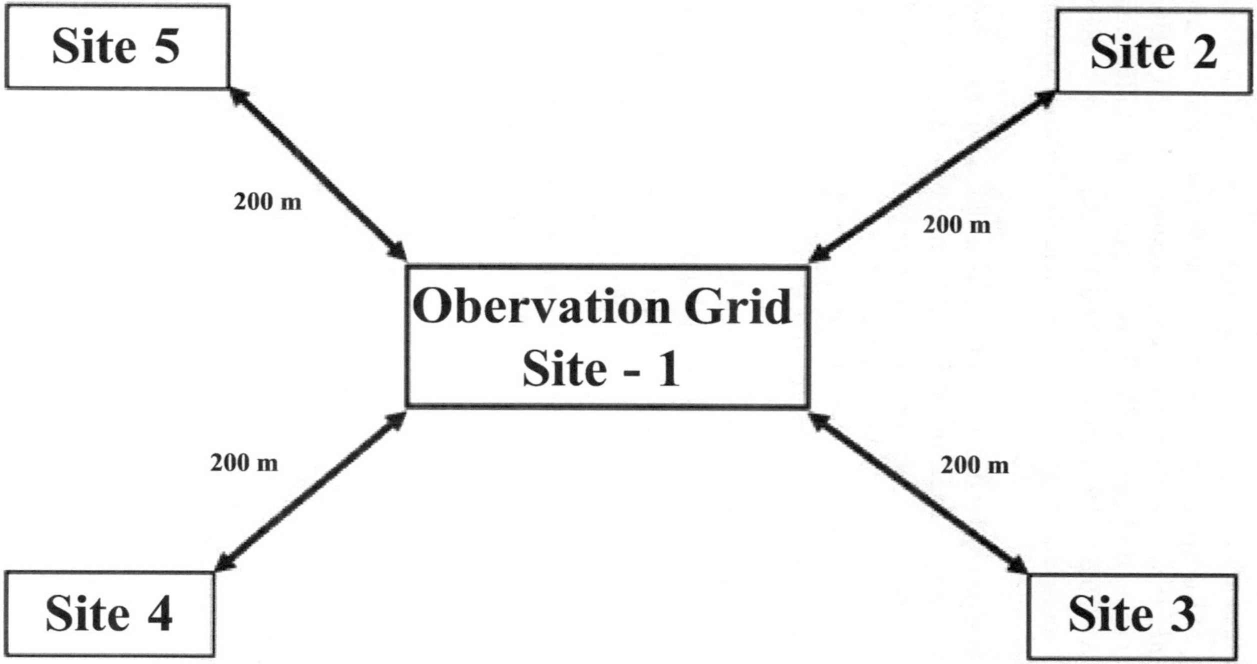


**Fig.1 :** Sampling scheme for collecting the composite samples.

## *Analysis of soil samples*

Surface soil samples from demarcated grid points and other related informations are collected through field survey. Soil samples are air dried, processed and analysed for pH, organic carbon, available phosphorous and potassium (Page *et al.*, 1982), available nitrogen (Subbaiah and Asija, 1956), available sulphur by using 0.15 percent $CaCl_2$ as the extractant (William and Steinbergs, 1959), available (DTPA extractable) Fe, Mn, Zn and Cu (Lindsay and Norvell, 1978) and available B (hot water soluble) by Carmine method (Hatcher and Wilcox, 1950).

The soils are grouped under different soil reaction classes viz extremely acidic (pH<4.5), very strongly acidic (pH 4.5 – 5.0 ), strongly acidic (pH 5.1 – 5.5), moderately acidic (pH 5.6-6.0), slightly acidic (pH 6.1-6.5), neutral (pH 6.6-7.3), slightly alkaline (pH 7.4-7.8), moderately alkaline (pH 7.9-8.4), strongly alkaline (pH 8.5-9.0) according to Soil Survey Manual (IARI, 1970).The soils are rated as low (below 0.50 %), medium (0.50-0.75 %) and high (above 0.75 %) in case of organic carbon, low (<280 kg $ha^{-1}$), medium (280 to 560 kg $ha^{-1}$) and high (>560 kg $ha^{-1}$) in case of available nitrogen, low (< 10 kg $ha^{-1}$), medium (10 to 25 kg $ha^{-1}$) and high (> 25 kg $ha^{-1}$) for available phosphorus, low (< 108 kg $ha^{-1}$), medium (108 to 280 kg $ha^{-1}$) and high (> 280 kg $ha^{-1}$) for available potassium and low (<10 mg $kg^{-1}$), medium (10-20 mg $kg^{-1}$) and high (> 20 mg $kg^{-1}$) for available sulphur (Singh *et. al.* 2004, Mehta *et. al.*1988). Critical limits of Fe, Mn, Zn, Cu and B, which separate deficient from non-deficient soils followed in India are 4.5, 2.0, 0.5, 0.2 and 0.5 mg $kg^{-1}$ respectively. (Follet and Lindsay, 1970 and Berger and Truog, 1940).

**Table 1 :** Methods used for analysis of soil samples

| Parameters | Methods | References |
|---|---|---|
| pH, EC | 1:2.5 (soil:water) | Jackson (1967) |
| Organic Carbon | Weight digestion (1N K2Cr2O7) | Walkley & Black. (1934) |
| Available N | Alkaline KMnO4 extractable -N | Subbiah & Asija (1956) |
| Olsen P | NaHCO3 extractable P (pH 8.5) | Olsen et al. (1954) |
| Bray P | NH4F extractable P | Bray et al. (1945) |
| Available K | NH4OAc extractable K | Jackson (1967) |
| Available S | 0.15% CaCl2 extractable S | Williams & Stienbergs (1969) |
| Available B | Hot water extractable B | Berger & Truog (1939) |
| Available Mo | Ammonium oxalate (pH 8.3) extractable Mo | Grig (1953) |
| Av. Fc, Mn, Zn, Cu | DTPA extractable | Lindsay & Norvel (1978) |

### *Organization of database in GIS*

The database consisting of geographic location, macro, major and micro-nutrients, texture, organic carbon, pH and management history for each grid points (**Geo-referenced**) are developed in **GIS**. The database has location co-ordinates of each geo-referenced sampling point (latitude and longitude), and also the name of village, block and district. The same is also structured in EXCEL format for statistical analysis.Macro, major and micro nutrients analysed for each grid points, are ranked as low, medium and high; micro-nutrients as sufficient and deficient; and pH class from extremely acidic to slightly alkaline based on the scheme.

### *Soil fertility maps*

Inverse weighted distance (IWD) a GIS based interpolation technique is used to develop individual maps of macro, major and micro nutrients of the study area. The area is calculated under different category of nutrients status. The maps for the above mentioned parameters have been prepared using Geographic Information System (GIS) from the data generated from analysis of grid soil samples.

Geo-referenced fertility maps of different nutrients are used to know the fertility status of farmers' fields even at the Village / Block / Tehsil/Mandal. It also provides suitable integrated nutrient fertilizer recommendation based on the resources available with the farmers. This approach helps in balanced fertilization and thereby improving the soil health and crop productivity. This information also will be useful for the planners, policy makers and administrators in the government, fertilizer manufacturers, distributors and extension workers of the Agriculture department/Horticulture department/NGOs.

## Case studies of JAMTARA district and JAMTARA block of Jharkhand state

### *Soil nutrition mapping at district level*

National Bureau of Soil Survey and Land Use Planning ( ICAR), Regional Centre, Kolkata in collaboration with the State department of Agriculture, Jharkhand state and the Department of Soil Science and Agricultural Chemistry, Birsa Agril. University, Ranchi undertook a project entitled "Assessment and mapping of some important soil parameters including soil acidity for the state of Jharkhand (1:50,000 scale) towards rational land use plan" . The major objectives of the project were :

- Preparation of districtwise soil acidity maps.

- Preparation of districtwise soil fertility maps (organic carbon, available N, P, K , S and available Fe, Mn, Zn, Cu and B).

The above maps will provide information regarding soil nutrients and soil acidity status for the districts, which will be very useful in identification of site specific problems for planning purposes.

## Methodology

The base map of Jamtara district was prepared on 1:50,000 scale using Survey of India toposheets (72L/8,12,16,72P/4,73I/9,13 and 73M/1,5) and all the maps were demarcated with grid points at 2.5 km interval.

The maps for the above mentioned parameters have been prepared using Geographic Information System (GIS) from data generated by analysis of grid soil samples.

## Soil Acidity and Fertility Status

### *Soil reaction*

Soil pH is an important property, which affects the availability of several plant nutrients. It is a measure of acidity and alkalinity and reflects the status of base saturation. The soils of the district have been grouped under seven soil reaction classes according to Soil Survey Manual (IARI, 1970).

The soil pH ranged from 4.3 to 7.4. The soil reaction classes with area are given in table 2 and figure 2. The data reveals that majority of the area is acidic (88 % of TGA), in which 36.6 per cent area is strongly acidic, 25.6 per cent very strongly acidic, 18.1 per cent moderately acidic, 5.6 per cent slightly acidic and 2.1 per cent extremely acidic in reaction. Soils of 5.4 per cent area of the district are neutral, whereas 3.3 per cent area is slightly alkaline in reaction.

**Table 2 :** Soils under different reaction classes

| Soil reaction | Area('00ha) | % of the TGA |
|---|---|---|
| Extremely acidic (pH 4.0 to 4.5) | 37 | 2.1 |
| Very strongly acidic (pH 4.6 to 5.0) | 462 | 25.6 |
| Strongly acidic (pH 5.1 to 5.5) | 659 | 36.6 |
| Moderately acidic (pH 5.6 to 6.0) | 327 | 18.1 |
| Slightly acidic (pH 6.1 to 6.5) | 101 | 5.6 |
| Neutral (pH 6.6 to 7.3) | 97 | 5.4 |
| Slightly alkaline (pH 7.4 to 7.8) | 60 | 3.3 |
| Miscellaneous | 59 | 3.3 |
| Total | 1802 | 100.0 |

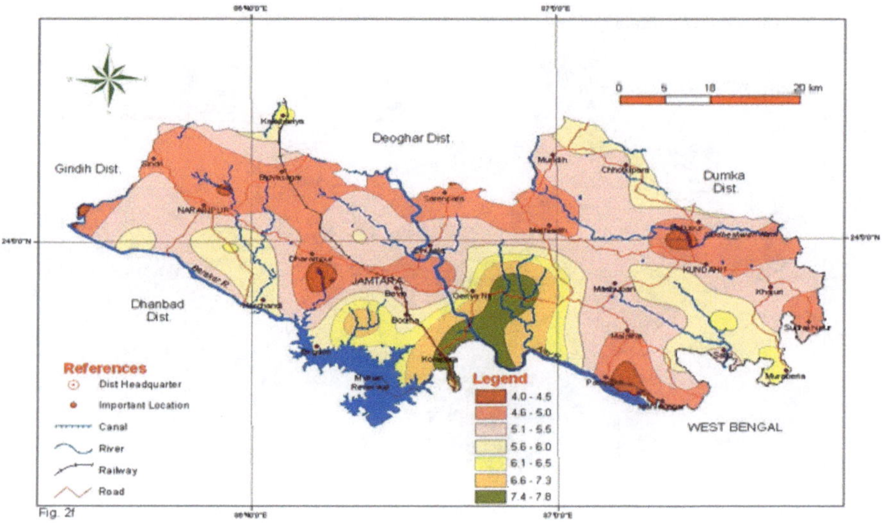


**Fig. 2 :** Soil reaction (pH) Jamtara district.

## *Organic Carbon*

The effect of soil organic matter on soil properties is well recognized. Soil organic matter plays a vital role in supplying plant nutrients, cation exchange capacity, improving soil aggregation and hence water retention and soil biological activity.

The organic carbon content in the district range from 0.16 to 1.47 %. They are mapped into three classes i.e., low (below 0.5 %), medium (0.5-0.75 %) and high (above 0.75 %) (Table 3 and Figure 3). From table 3, it is seen that 47.4 per cent area of the district shows high organic carbon content. Medium and low organic carbon content constitute 24.7 and 24.6 per cent area, respectively.

**Table 3 :** Organic carbon status

| Organic carbon (%) | Area ('00ha) | % of the TGA |
|---|---|---|
| Low (below 0.50 %) | 444 | 24.6 |
| Medium (0.50-0.75 %) | 445 | 24.7 |
| High (above 0.75 %) | 854 | 47.4 |
| Miscellaneous | 59 | 3.3 |
| Total | 1802 | 100.0 |

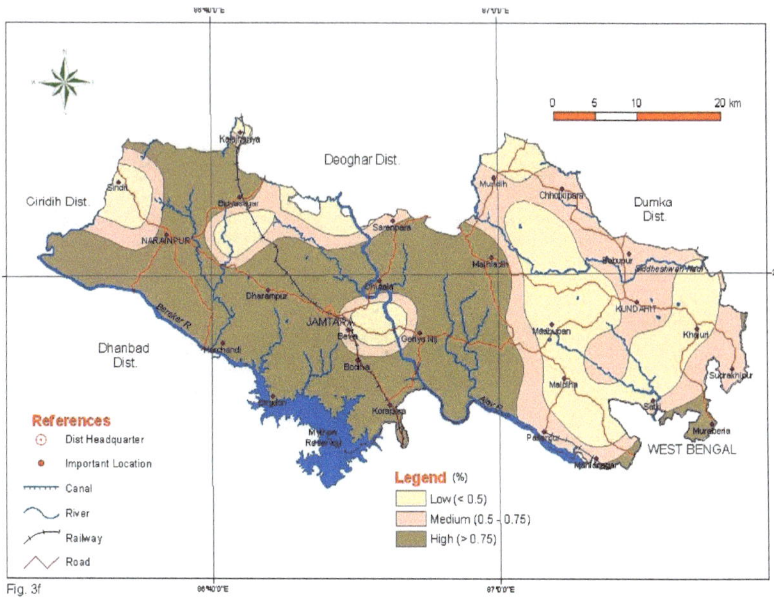


**Fig. 3 :** Organic Carbon-Jamtara district Jharkhand

## *Macro Nutrients*

Nutrients, like nitrogen (N), phosphorus (P) and potassium (K) are considered as primary nutrients and sulphur (S) as secondary nutrient. These nutrients help in proper growth, development and yield differentiation of plants and are generally required by plants in large quantity.

## *Available Nitrogen*

Nitrogen is an integral component of many compounds including chlorophyll and enzyme essential for plant growth. It is an essential constituent for amino acids, which is building blocks for plant tissue, cell nuclei and protoplasm. It encourages above ground vegetative growth and deep green colour to leaves. Deficiency of nitrogen decreases rate and extent of protein synthesis and result into stunted growth and develop chlorosis.

Available nitrogen content in the surface soils of the district ranges between 135 and 543 kg/ha and details are given in table 4 and figure 4. Majority of the soils (76.5 % of TGA) of the district have medium availability status of nitrogen (280-560 kg $ha^{-1}$) and soils of 20.2 per cent area have low available nitrogen content (<280 kg $ha^{-1}$).

**Table 4 :** Available nitrogen status in the surface soils

| Available nitrogen (kg ha$^{-1}$) | Area('00ha) | % of the TGA |
|---|---|---|
| Low (below 280) | 365 | 20.2 |
| Medium (280-560) | 1378 | 76.5 |
| Miscellaneous | 59 | 3.3 |
| Total | 1802 | 100.0 |

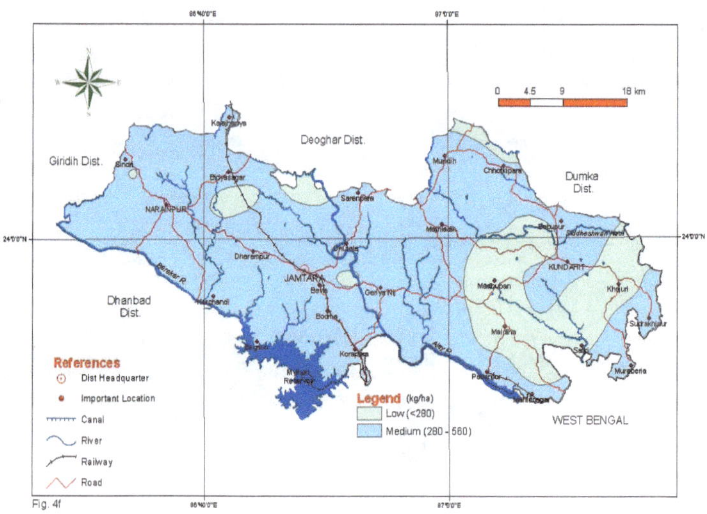


**Fig. 4 :** Available nitrogen - Jamtara district, Jharkhand.

## *Available Phosphorus*

Phosphorus is important component of adenosine di-phosphate (ADP) and adenosine tri-phosphate (ATP), which are involved in energy transformation in plant. It is essential component of deoxyribonucleic acid (DNA), the seat of genetic inheritance in plant and animal. Phosphorous takes part in important functions like photosynthesis, nitrogen fixation, crop maturation, root development, strengthening straw in cereal crops etc. The availability of phosphorous is restricted under acidic and alkaline soil reaction, mainly due to P-fixation. In acidic condition, it get fixed with aluminum and iron and in alkaline condition with calcium.

Available phosphorus content in these soils ranges between 1.4 and 18.5 kg/ha and their distribution is given in table 5 and figure 5. Data reveals that majority of the soils are low (73.1 % of TGA) followed by medium (23.6 % of TGA) content of available phosphorous.

**Table 5 :** Available phosphorous status in the surface soils

| Available phosphorous (kg ha$^{-1}$) | Area('00ha) | % of the TGA |
|---|---|---|
| Low (below 10) | 1318 | 73.1 |
| Medium (10-25) | 425 | 23.6 |
| Miscellaneous | 59 | 3.3 |
| Total | 1802 | 100.0 |

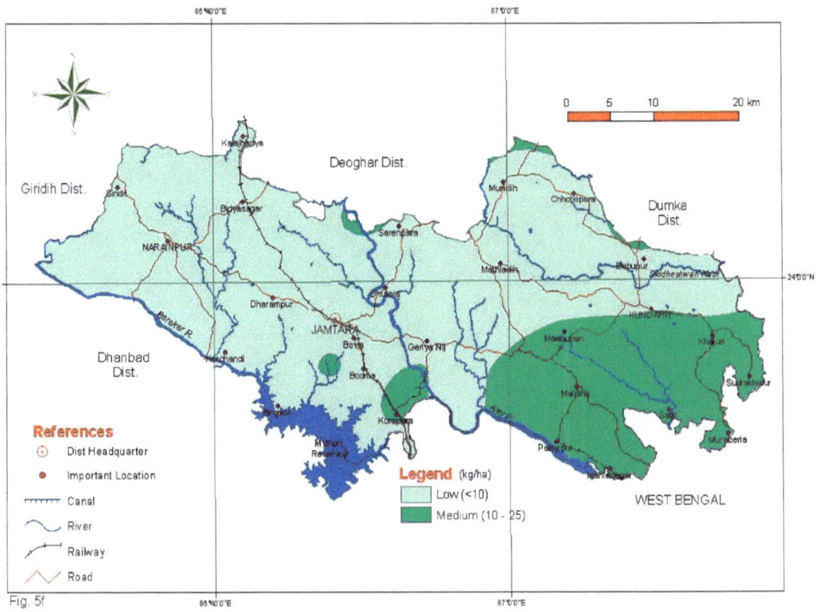


**Fig. 5 :** Available Phosphorus-Jamtara District Jharkhand

## *Available Potassium*

Potassium is an activator of various enzymes, responsible for plant processes like energy metabolism, starch synthesis, nitrate reduction and sugar degradation. It is extremely mobile in plant and help to regulate opening and closing of stomata in the leaves and uptake of water by root cells. It is important in grain formation and tuber development and encourages crop resistance for certain fungal and bacterial diseases.

Available potassium content in these soils ranges between 39 and 582 kg/ ha and details about area and distribution is given in table 6 and figure 6. The data reveal that most of the soils (61.3 % of TGA) have medium available potassium content (108-280 kg ha$^{-1}$). Soils of 14.1 per cent area are high (above 280 kg ha$^{-1}$) and 21.3 per cent area are low (below 108) in available potassium content.

**Table 6 :** Available potassium status in the surface soils

| Available potassium (kg ha$^{-1}$) | Area('00ha) | % of the TGA |
|---|---|---|
| Low (below 108) | 383 | 21.3 |
| Medium (108-280) | 1105 | 61.3 |
| High (above 280) | 255 | 14.1 |
| Miscellaneous | 59 | 3.3 |
| Total | 1802 | 100.0 |

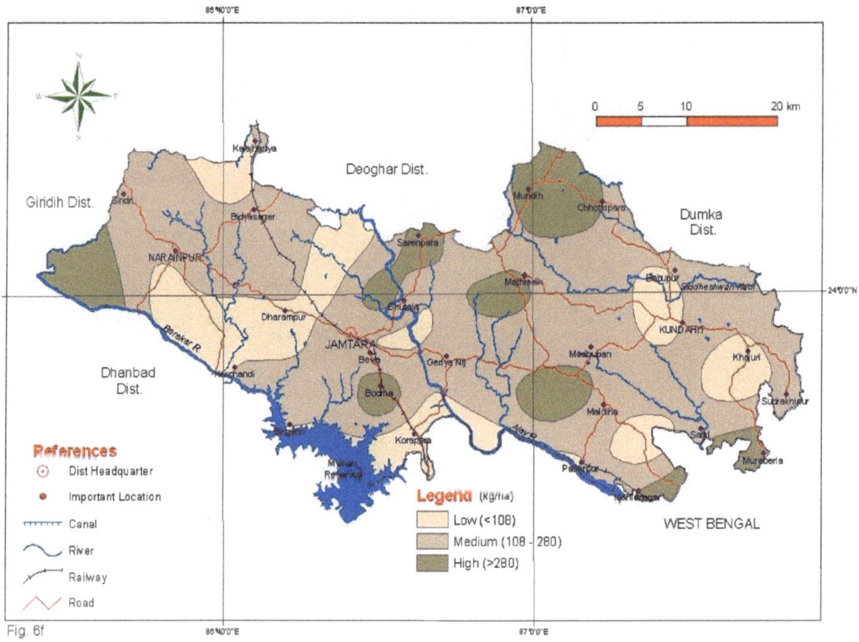


**Fig. 6 :** Available Potassium-Jamtara District, Jharkhand.

### *Available Sulphur*

Sulphur is essential in synthesis of sulphur containing amino acids (cystine, cysteine and methionine), chlorophyll and metabolites including co-enzyme A, biotin, thiamine, or vitamin B1 and glutathione. It activates many proteolytic enzymes, increase root growth and nodule formation and stimulate seed formation.

The available sulphur content in the soils ranges from 0.76 to 41.49 mg kg$^{1}$ and details about area and distribution is given in table 7 and figure 7. Soils of 25.8 per cent of the area are low (<10 mg kg$^{-1}$) whereas soils of 39.7 and 31.2 per cent area are medium (10-20 mg kg$^{-1}$) and high (>20 mg kg$^{-1}$) in available sulphur content, respectively.

**Table 7 :** Available sulphur status in the surface soils

| Available sulphur(mg $kg^{-1}$) | Area('00ha) | % of the TGA |
|---|---|---|
| Low (<10) | 465 | 25.8 |
| Medium (10-20) | 715 | 39.7 |
| High (>20) | 563 | 31.2 |
| Miscellaneous | 59 | 3.3 |
| Total | 1802 | 100.0 |

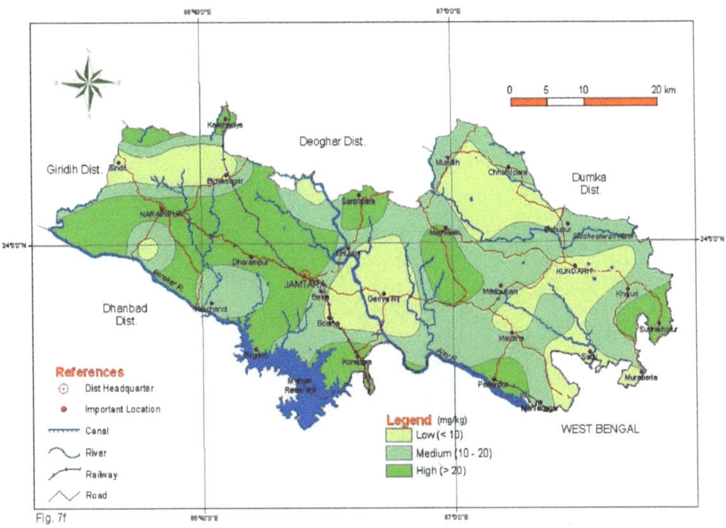


**Fig. 7 :** Availlable Sulphur Jamtara district, Jharkhand.

## *Micronutrients*

Proper understanding of micronutrients availability in soils and extent of their deficiencies is a pre-requisite for efficient management of micronutrient fertilizers to sustain crop productivity. Therefore, it is essential to know the micronutrients status of soil, before introducing any type of land use.

### *Available Iron*

Iron is constituent of cytochromes, several enzymes. It is capable of acting as electron carrier in many enzyme systems that brings about oxidation-reduction reactions in plants. It promotes starch formation and seed maturation.

The available iron content in the surface soils ranges between 5.9 and 82.0 mg $kg^{-1}$. As per the critical limit of available iron (> 4.5 mg $kg^{-1}$), all the soils are sufficient in available iron. They are grouped and mapped into four classes. Majority of the soils (50.6 % of TGA) have available iron content between the ranges of 50 to 100 mg $kg^{-1}$. The details of area and distribution are presented in table 8 and figure 8.

**Table 8 :** Available iron status in the surface soils

| Available iron (mg $kg^{-1}$) | Area('00ha) | % of the TGA | Rating |
|---|---|---|---|
| <15 | 116 | 6.4 | Sufficient |
| 15-25 | 153 | 8.5 | |
| 25-50 | 562 | 31.2 | |
| 50-100 | 912 | 50.6 | |
| Miscellaneous | 59 | 3.3 | |
| Total | 1802 | 100.0 | |

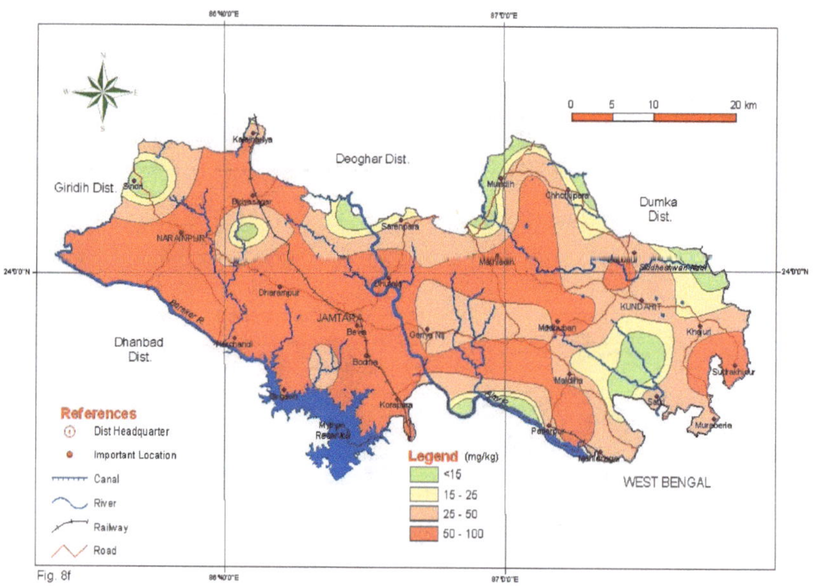


**Fig. 8 :** Available Iron Jamtara District, Jharkhand.

## *Available Manganese*

Manganese is essential in photosynthesis and nitrogen transformations in plants. It activates decarboxylase, dehydrogenase, and oxidase enzymes.

The available manganese content in surface soils ranges between 5.9 and 83.2 mg $kg^{-1}$. As per the critical limit of available manganese (> 2 mg $kg^{-1}$), all the soils are sufficient in available manganese. They are grouped and mapped into four classes. Most of soils (44.5 %) have available Mn content between 10 and 25 mg $kg^{-1}$. The details of area and distribution are presented in Table 9 and Figure 9.

**Table 9 :** Available manganese status in the surface soils

| Available manganese(mg $kg^{-1}$) | Area('00ha) | % of the TGA | Rating |
|---|---|---|---|
| <10 | 150 | 8.3 | |
| 10-25 | 802 | 44.5 | |
| 25-50 | 273 | 15.2 | Sufficient |
| 50-100 | 518 | 28.7 | |
| Miscellaneous | 59 | 3.3 | |
| Total | 1802 | 100.0 | |

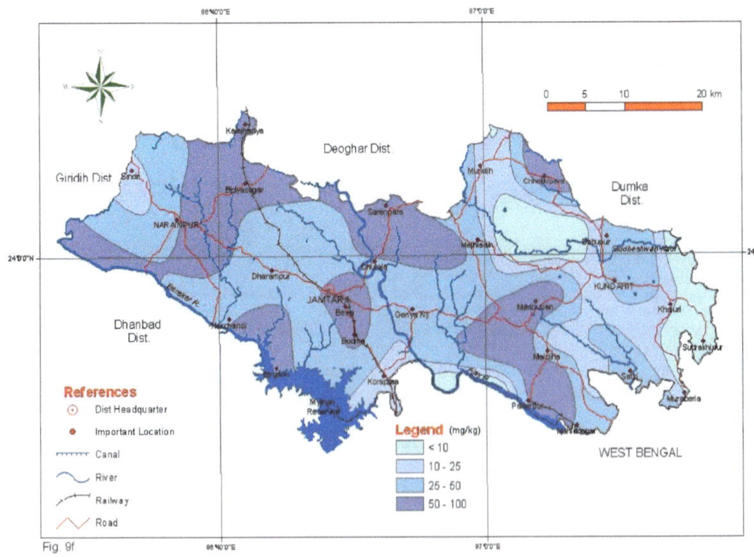


**Fig. 9 :** Available Manganese, Jamtar District, Jharkhand.

### *Available Zinc*

Zinc plays role in protein synthesis, reproductive process of certain plants and in the formation starch and some growth hormones. It promotes seed maturation and production.

The available zinc in surface soils ranges between 0.38 and 5.00 mg $kg^{-1}$. They are grouped and mapped into five classes. Majority of soils (92.9 % of TGA) are sufficient (>0.5 mg $kg^{-1}$), whereas soils of 3.8 per cent area are deficient (<0.5 mg $kg^{-1}$) in available zinc. The details of area and distribution are presented in Table 10 and Figure 10).

**Table 10 :** Available zinc status in the surface soils

| Available zinc(mg kg-1) | Area ('00ha) | % of the TGA | Rating |
|---|---|---|---|
| <0.5 | 68 | 3.8 | Deficient |
| 0.5-1.0 | 303 | 16.8 | Sufficient |
| 1.0-2.0 | 1011 | 56.1 | |
| 2.0-3.0 | 253 | 14.0 | |
| 3.0-5.0 | 108 | 6.0 | |
| Miscellaneous | 59 | 3.3 | |
| Total | 1802 | 100.0 | |

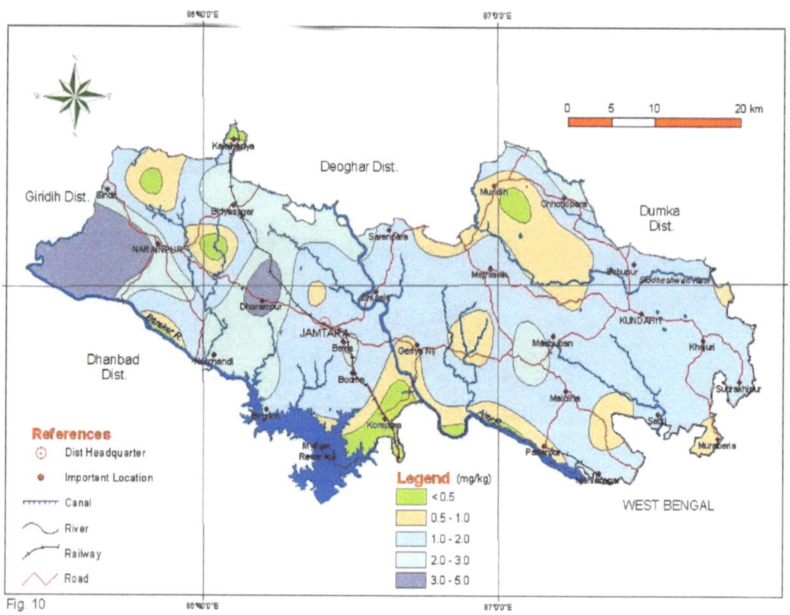


**Fig. 10 :** Available Zinc Jamtara district, Jharkhand

## *Available Copper*

Copper is involved in photosynthesis, respiration, protein and carbohydrate metabolism and in the use of iron. It stimulates lignifications of all the plant cell wall and is capable of acting as electron carrier in many enzyme systems, that bring about oxidation-reduction reactions in plants.

The available copper status in surface soils ranges between 0.14 and 5.06 mg $kg^{-1}$. They are grouped and mapped into six classes. Majority of soils (95.8 % of TGA) have sufficient amount of available copper (>0.2 mg $kg^{-1}$) and soils of 0.9 % area are deficient in available copper (<0.2 mg $kg^{-1}$). The details of area and distribution are presented in (Table 11 and Figure 11).

**Table 11 :** Available copper status in the surface soils

| Available copper(mg $kg^{-1}$) | Area('00ha) | % of the TGA | Rating |
|---|---|---|---|
| <0.2 | 16 | 0.9 | Deficient |
| 0.2-0.5 | 50 | 2.8 | |
| 0.5-1.0 | 58 | 3.2 | |
| 1.0-2.0 | 290 | 16.1 | Sufficient |
| 2.0-4.0 | 1208 | 67.0 | |
| 4.0-6.0 | 121 | 6.7 | |
| Miscellaneous | 59 | 3.3 | |
| Total | 1802 | 100.0 | |

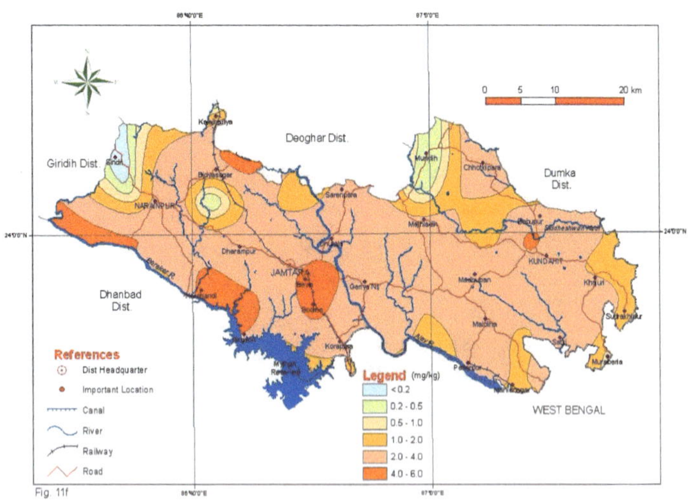


**Fig. 11 :** Available copper Jamtara district, Jharkhand.

## *Available Boron*

Boron increases solubility and mobility of calcium in plants and it act as regulator of K/Ca ratio in the plant. It is required for development of new meristematic tissue and also necessary for proper pollination, fruit and seed setting and translocation of sugar, starch and phosphorous etc. It has role in synthesis of amino acid and protein and regulates carbohydrate metabolism.

The available boron content in the soils ranges from 0.02 to 6.09 mgkg$^{-1}$ and details about area and distribution is given in table 12 and figure 12. The critical limit for deficiency of the available boron is <0.5. Soils of 23.0 per cent area of district are deficient (<0.50 mgkg$^{-1}$) whereas 73.7 per cent area are sufficient (>0.50 mgkg$^{-1}$) in available boron content.

**Table 12 :** Available boron status in the surface soils

| Available boron(mg kg-1) | Area('00ha) | % of the TGA | Rating |
|---|---|---|---|
| <0.25 | 290 | 16.1 | Deficient |
| 0.25-0.50 | 125 | 6.9 | |
| 0.50-0.75 | 169 | 9.4 | Sufficient |
| >0.75 | 1159 | 64.3 | |
| Miscellaneous | 59 | 3.3 | |
| Total | 1802 | 100.0 | |

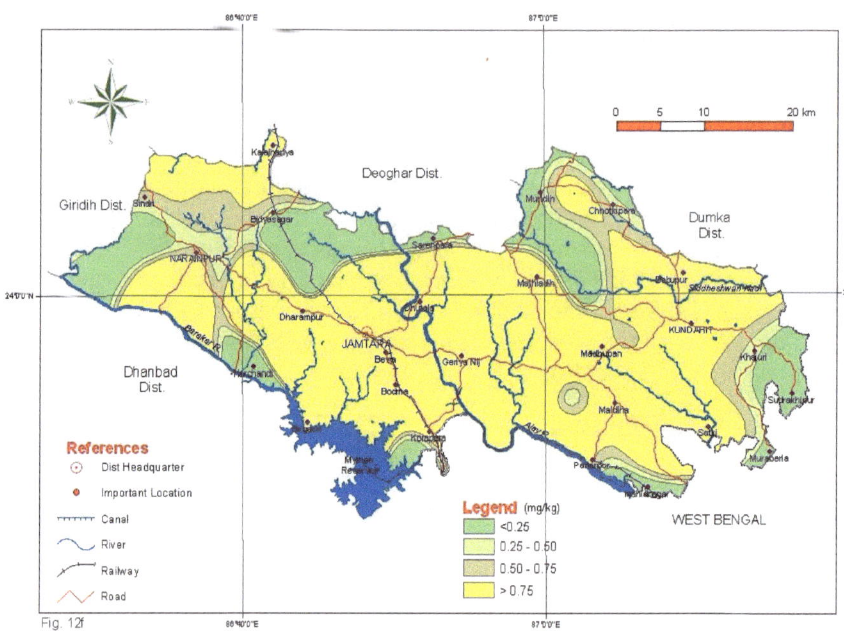


**Fig. 12 :** Available boron Jamtara District, Jharkhand.

## Summary

The soil pH ranges from 4.3 to 7.4. Soils of 88 % area are acidic in reaction, in which 36.6 per cent area strongly acidic, 25.6 per cent very strongly acidic, 18.1 per cent moderately acidic, 5.6 per cent slightly acidic and 2.1 per cent, extremely acidic in reaction. Soils of 5.4 per cent area of the district are neutral whereas 3.3 per cent area is slightly alkaline in reaction. The organic carbon content in the district ranges from 0.16 to 1.47 %. Soils of 47.4 per cent area of the district shows high organic carbon content. Medium and low organic carbon content constitute 24.7 and 24.6 per cent area, respectively.

Available nitrogen content in the surface soils of the district ranges between 135 and 543 kg/ha. Majority of the soils (comprising low lands/medium lands) of the district have medium availability of nitrogen and soils of 20.2 per cent area have low available nitrogen content. Available phosphorus content in these soils ranges between 1.4 and 18.5 kg/ha. Majority of the soils are low (73.1 % of TGA), mainly because of soil acidity problems followed by soils with medium (23.6 % of TGA) category of available phosphorous. Available potassium content in these soils ranges between 39 and 582 kg/ha. Most of the soils (61.3 % of TGA) have medium available potassium content. Soils of 14.1 per cent area have high and 21.3 per cent area have low in available potassium . The available sulphur content in the soils ranges from 0.76 to 41.49 mg $kg^{-1}$. Soils of 25.8 per cent of the area are low (<10 mg $kg^{-1}$), whereas soils of 39.7 and 31.2 per cent area are medium (10-20 mg $kg^{-1}$) and high (>20 mg $kg^{-1}$) in available sulphur , respectively.

Soils we analysed for available (DTPA extractable) micronutrients and it is evident that, all the soils are sufficient in available iron and manganese, whereas soils of 3.8 and 0.9 per cent area are deficient in available zinc and copper, respectively. Available boron content in the soils ranges between 0.02 to 6.09 mg$kg^{-1}$and 23.0 per cent area of district is deficient (<0.50 mg $kg^{-1}$).

Results of the study point out the need to ameliorate acidic soils, to enhance the productivity of major crops, such as pulses, oilseeds, maize etc.Further, combined application of well decomposed organic manures along with balanced dose of NPK fertilisers is required to increase production and sustain soil health. Among the secondary and micronutrients, Sulphur and Boron deficiency is common in 1 out of 4 soils.This is important,especially for yield and quality of oilseeds,pulses and vegetable crops grown in large parts of the district. Soil testing laboratories of the region must take note of these two nutrients for long term sustenance of soil health.

### *Soil Nutrient Mapping at Block level*

Soil nutrient mapping at district level was attempted by collecting soil samples at the interval of 2.5 km. (NBSS & LUP 2006). The resultant map could be utilized for district planning. However, for translating the map information for farm planning, soil sampling at closer interval (500 meter) is appropriate, to capture all kinds of variability in nutrient status. For bridging the outlined gaps, NBSS & LUP, Nagpur and Regional Centre, Kolkata in collaboration with the

Department of Soil Science and Agricultural Chemistry, BAU, Ranchi and Department of Agriculture & Cane Development, Govt. of Jharkhand, has taken up model project in Jharkhand state entitled **"Assessment and mapping of some important soil parameters including macro and micro nutrients at block level of Dumka, Jamtara, Hazaribagh and Ramgarh districts for optimum land use plan".**

### *Objectives*

The major objectives of the project were

- Preparation of GIS aided block wise soil acidity maps.
- Preparation of block wise soil nutrient maps (organic carbon, available P, K, S and available Fe, Mn, Zn, Cu, B).

The above maps will provide information regarding soil nutrients and soil acidity status for different blocks in the district, which will help identify site specific problems in different blocks, for need based plan and implementation.

## Methodology

The base map of all the blocks was prepared on 1:50,000 scale using Survey of India toposheets (72L/8,12,16,72P/4,73I/9,13 and 73M/1,5) and PS/ Block maps of all the blocks and maps were demarcated with grid points at 500m interval.

Seven thousand two hundred nine (7209) composite surface soil samples at 0-25 cm depth from demarcated grid points in six blocks of Jamtara district (Table 13 ) and other related information were collected through field survey.

**Table 13 :** No of collected surface soil samples in different blocks

| Blocks | Area ( sq.km) | Surface soil samples |
|---|---|---|
| Fatehpur | 263.7 | 1064 |
| Jamtara | 352.1 | 1277 |
| Karmatarn | 177.1 | 712 |
| Kundahit | 296.5 | 1231 |
| Nala | 440.0 | 1740 |
| Narayanpur | 283.6 | 1185 |
| Total | 7209 | |

The maps for the different soil parameters have been prepared using Geographic Information System (GIS) from data generated by analysis of grid soil samples.

## Case Study in Kundahit Block of Jamtara District

### *Soil Texture*

Coarse loamy soils with sandy loam textural class and sandy soils with loamy sand textural class together cover major area (69.5% of TGA) in the block. Fine loamy soils with loam, sandy clay loam, clay loam textural classes occupy 29.2 per cent area. (Table 13a & Figure 13).

**Table 13a :** Soil textural class

| Textural class (Surface) | Particle size class (Textural grouping) | Area($km^2$) | % Area |
|---|---|---|---|
| Loamy sand | Sandy | 87.4 | 29.5 |
| Sandy loam | Coarse loamy | 118.6 | 40.0 |
| Loam | | 63.4 | 21.4 |
| Sandy clay loam | Fine loamy | 17.7 | 5.9 |
| Clay loam | | 5.5 | 1.9 |
| **Miscellaneous** | | **3.9** | **1.3** |
| **Total** | | **296.5** | **100.0** |

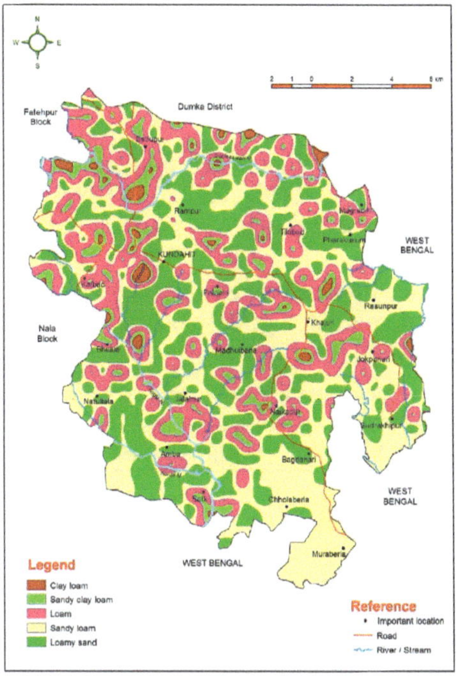


**Fig. 13 :** Surface Texture Kundahit Block, Jamtara district, Jharkhand.

## *Soil Reaction*

The soil pH ranges from 4.3 to 7.2. The soil reaction class with area are given in (Table 14 and Figure 14). The data reveal that majority of the area is acidic (91.4% of TGA), in which 30.8 per cent area is strongly acidic, 18.0 per cent very strongly acidic, 22.6 per cent moderately acidic, 16.0 per cent slightly acidic and 4.0 per cent extremely acidic in reaction. Soils of 7.3 per cent area of the block are neutral.

**Table 14 :** Soils under different reaction classes

| Soil reaction | Area ('00ha) | % of the TGA |
|---|---|---|
| Extremely acidic (pH 4.0 to 4.5) | 11.9 | 4.0 |
| Very strongly acidic (pH 4.6 to 5.0) | 53.4 | 18.0 |
| Strongly acidic (pH 5.1 to 5.5) | 91.3 | 30.8 |
| Moderately acidic (pH 5.6 to 6.0) | 67.0 | 22.6 |
| Slightly acidic (pH 6.1 to 6.5) | 47.4 | 16.0 |
| Neutral (pH 6.6 to 7.3) | 21.6 | 7.3 |
| Miscellaneous | 3.9 | 1.3 |
| Total | 296.5 | 100.0 |

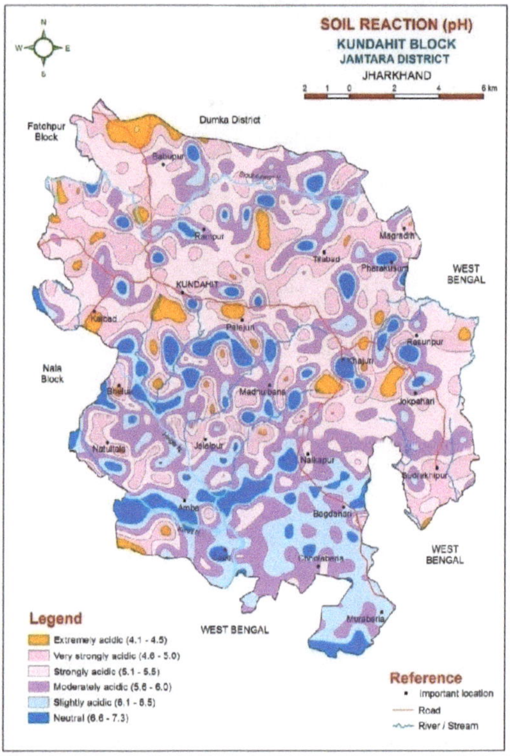


**Fig. 14 :** Soil reaction (pH) Kundahit block Jamtara district, Jharkhand.

## *Organic Carbon*

The organic carbon content in the block ranges from 0.15 to 0.71 percent.They are mapped into two classes i.e., low (below 0.5%) and medium (0.5-0.75%) (Table 15 and Figure 15) From table 15, it is seen that 72.8 per cent area of the block are low in organic carbon content and medium organic carbon content are found in 25.9 per cent area of the block.

**Table 15 :** Organic carbon status

| Organic carbon (%) | Area('00ha) | % of the TGA |
|---|---|---|
| Low (below 0.50%) | 215.7 | 72.8 |
| Medium (0.50-0.75%) | 76.9 | 25.9 |
| Miscellaneous | 3.9 | 1.3 |
| Total | 296.5 | 100.0 |

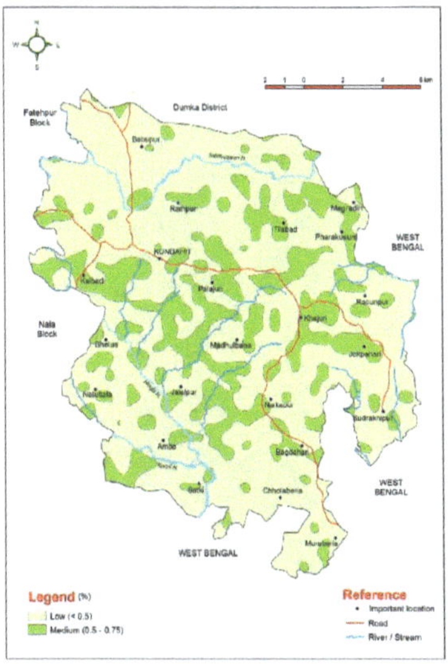


**Fig. 15 :** Organic Carbon, Kundahit block, Jamtara district, Jharkhand.

## *Macronutrients*

Nutrients like phosphorus (P) and potassium (K) are considered as primary nutrients and sulphur (S) as secondary nutrient. These nutrients help in proper growth, development and yield differentiation of plants and are generally required by plants in large quantity.

### *Available Phosphorus*

Available phosphorus content in these soils ranges between 1.7 and 34.4 kg $ha^{-1}$and their distribution is given in (Table 16 and Figure 16). Data reveal that 50.8 per cent area of the block are low in available phosphorous content and 43.2 per cent area of the block are medium content of available phosphorous.

**Table 16 :** Available phosphorous status in the surface soils

| Available phosphorous (kg ha$^{-1}$) | Area('00ha) | % of the TGA |
|---|---|---|
| Low (below 10) | 150.7 | 50.8 |
| Medium (10-25) | 128.0 | 43.2 |
| High (above 25) | 13.9 | 4.7 |
| Miscellaneous | 3.9 | 1.3 |
| Total | 296.5 | 100.0 |

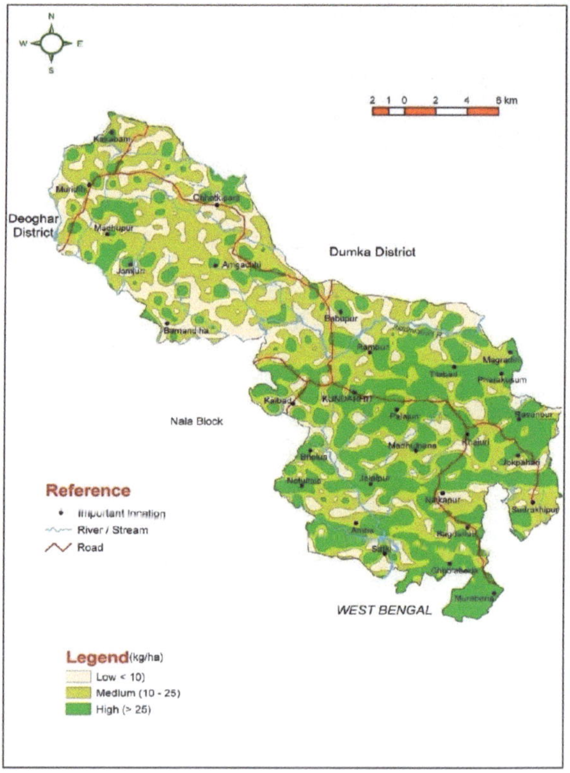


**Fig. 16 :** Available phosphorous Kundahit block, Jamtara district, Jharkhand.

## *Available Potassium*

Available potassium content in these soils ranges between 64.8 and 612.3 kg ha$^{-1}$and details about area and distribution is given in (Table 17 and Figure 17). The data reveal that most of the soils (47.6% of TGA) have medium available potassium content (108-280 kg ha$^{-1}$). Soils of 26.7 per cent area are high (above 280 kg ha$^{-1}$) and 24.4 per cent area are low (below 108 kg ha$^{-1}$) in available potassium content.

**Table 17 :** Available potassium status in the surface soils

| Available potassium (kg ha$^{-1}$) | Area('00ha) | % of the TGA |
|---|---|---|
| Low (below 108) | 72.2 | 24.4 |
| Medium (108-280) | 141.2 | 47.6 |
| High (above 280) | 79.2 | 26.7 |
| Miscellaneous | 3.9 | 1.3 |
| Total | 296.5 | 100.0 |

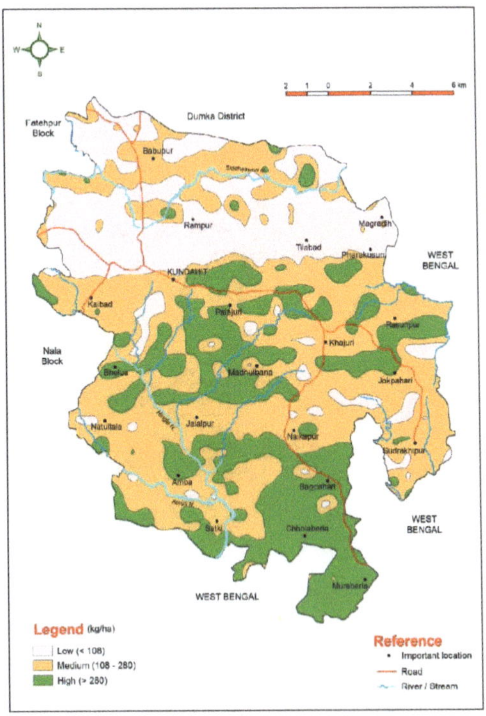


**Fig. 17 :** Available potassium, Kundahit block, Jamtara district, Jharkhand.

### *Available Sulphur*

The available sulphur content in the soils ranges from 6.2 to 52.3 mg kg$^{-1}$ and details about area and distribution is given in (Table 18). Soils of 13.7 per cent of the area are low (<10 mg kg$^{-1}$) whereas soils of 69.1 and 15.9 per cent area are medium (10-20 mg kg$^{-1}$) and high (>20 mg kg$^{-1}$) in available sulphur content, respectively.

**Table 18 :** Available sulphur status in the surface soils

| Available sulphur (mg $kg^{-1}$) | Area('00ha) | % of the TGA |
|---|---|---|
| Low (<10) | 40.6 | 13.7 |
| Medium (10-20) | 204.9 | 69.1 |
| High (>20) | 47.1 | 15.9 |
| Miscellaneous | 3.9 | 1.3 |
| Total | 296.5 | 100.0 |

### *Available Iron*

The available iron content in the surface soils ranges between 5.9 and 92.0 mg $kg^{-1}$. As per the critical limit of available iron (<4.5 mg $kg^{-1}$), all the soils of the block are sufficient in available iron. They are grouped and mapped into four classes. Majority of the soils (54.3% of TGA) have available iron content between the range of 25 to 50 mg $kg^{-1}$. The details of area and distribution are presented in (Table 19).

**Table 19 :** Available iron status in the surface soils

| Available iron (mg $kg^{-1}$) | Area('00ha) | % of the TGA | Rating |
|---|---|---|---|
| 4.5 - 10 | 13.8 | 4.7 | Sufficient |
| 10 - 25 | 100.1 | 33.8 | |
| 25 - 50 | 161.1 | 54.3 | |
| 50 - 100 | 17.6 | 5.9 | |
| Miscellaneous | 3.9 | 1.3 | |
| Total | 296.5 | 100.0 | |

### *Available Manganese*

The available manganese content in surface soils ranges between 4.4 and 59.3 mg $kg^{-1}$. As per the critical limit of available manganese (<2 mg $kg^{-1}$), all the soils of the block are sufficient in available manganese. They are grouped and mapped into four classes. Most of soils (55.7%) have available Mn content between 25 and 50 mg $kg^{-1}$. The details of area and distribution are presented in (Table 20).

**Table 20 :** Available manganese status in the surface soils

| Available manganese(mg $kg^{-1}$) | Area('00ha) | % of the TGA | Rating |
|---|---|---|---|
| 2-10 | 8.6 | 2.9 | Sufficient |
| 10-25 | 116.4 | 39.3 | |
| 25-50 | 165.3 | 55.7 | |
| 50-100 | 2.3 | 0.8 | |
| Miscellaneous | 3.9 | 1.3 | |
| Total | 296.5 | 100.0 | |

### *Available Zinc*

The available zinc in surface soils ranges between 0.24 and 6.75 mg $kg^{-1}$. They are grouped and mapped into six classes.Majority of soils (89.3% of TGA) are sufficient (>0.5 mg $kg^{-1}$), whereas soils of 9.4 per cent area are deficient (<0.5 mg $kg^{-1}$) in available zinc. The details of area and distribution are presented in (Table 21 and Figure 18).

**Table 21 :** Available zinc status in the surface soils

| Available zinc(mg $kg^{-1}$) | Area('00ha) | % of the TGA | Rating |
|---|---|---|---|
| <0.5 | 28.0 | 9.4 | Deficient |
| 0.5-1.0 | 86.8 | 29.3 | |
| 1.0-2.0 | 116.5 | 39.3 | |
| 2.0-3.0 | 41.9 | 14.2 | |
| 3.0-5.0 | 16.3 | 5.5 | |
| >5.0 | 3.1 | 1.0 | |
| Miscellaneous | 3.9 | 1.3 | |
| Total | 296.5 | 100.0 | |

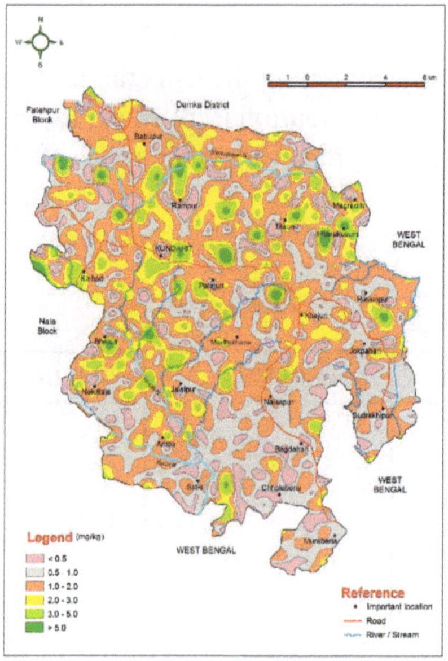


**Fig. 18 :** Available Zinc, Kundahit block, Jamtara district.

## *Available Copper*

The available copper status in surface soils ranges between 0.04 and 7.18 mg $kg^{-1}$. They are grouped and mapped into six classes. Majority of soils (95.8% of TGA) have sufficient amount of available copper (>0.2 mg $kg^{-1}$) and soils of 2.9 per cent area are deficient in available copper (<0.2 mg $kg^{-1}$). The details of area and distribution are presented in (Table 22).

**Table 22 :** Available copper status in the surface soils

| Available copper(mg $kg^{-1}$) | Area('00ha) | % of the TGA | Rating |
|---|---|---|---|
| <0.2 | 8.4 | 2.9 | Deficient |
| 0.2-0.5 | 20.2 | 6.8 | |
| 0.5-1.0 | 64.3 | 21.7 | |
| 1.0-2.0 | 129.4 | 43.6 | |
| 2.0-4.0 | 67.6 | 22.8 | |
| >4.0 | 2.7 | 0.9 | |
| Miscellaneous | 3.9 | 1.3 | |
| Total | 296.5 | 100.0 | |

### *Available Boron*

The available boron content in the soils ranges from 0.02 to 4.93 mg $kg^{-1}$ and details about area and distribution is given in (Table 23 and Figure 19). The critical limit for deficiency of the available boron is <0.50 mg $kg^{-1}$. Soils of 24.6 per cent area of block are deficient (<0.50 $mgkg^{-1}$) whereas 74.1 per cent area are sufficient (>0.50 $mgkg^{-1}$) in available boron content.

**Table 23 :** Available boron status in the surface soils

| Available boron(mg $kg^{-1}$) | Area ('00ha) | % of the TGA | Rating |
|---|---|---|---|
| <0.50 | 72.9 | 24.6 | Deficient |
| 0.50-0.75 | 92.0 | 31.1 | |
| 0.75-1.00 | 62.7 | 21.1 | Sufficient |
| >1.00 | 65.0 | 21.9 | |
| Miscellaneous | 3.9 | 1.3 | |
| Total | 296.5 | 100.0 | |

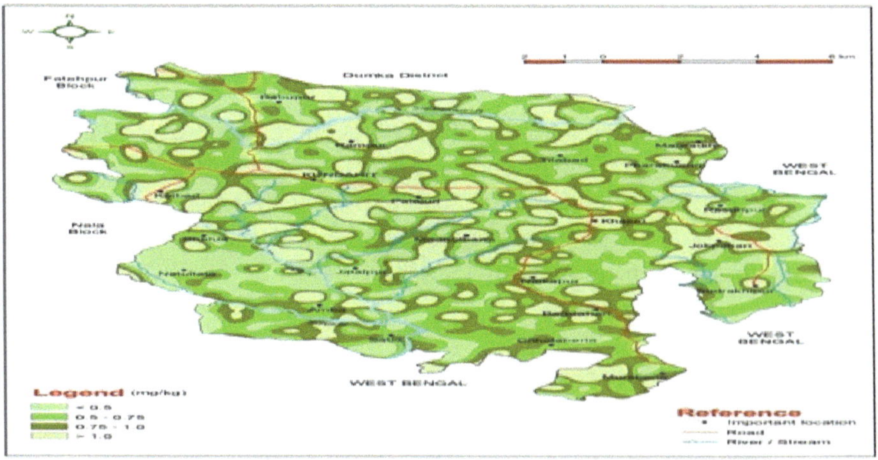


**Fig. 19 :** Available boron, Kundahit block, Jamtara district, Jharkhand.

## Summary

Coarse loamy and sandy surface soils in the Kundahit block cover 40.0 per cent and 29.2 per cent area, respectively. Soils of 91.4 per cent area are acidic in reaction, out of which 52.8 per cent area is strongly to extremely acidic (pH 4.0 to 5.5) and 38.6 per cent area is slightly to moderately acidic (pH 5.6 to 6.5) in reaction. Soils of 7.3 per cent area of the block are neutral.

The organic carbon content in the block ranges from 0.15 to 0.71per cent and 72.8 per cent area of the block are low in organic carbon content. Soils of 50.8 per cent area of the block are low in available phosphorous content. Available potassium content in these soils ranges between 64.8 and 612.3 kg $ha^{-1}$and 24.4 per cent area are low in available potassium content. The available sulphur content in the soils varies from 6.2 to 52.3 mg $kg^{-1}$. Soils of 13.7 per cent of the area are low (<10 mg $kg^{-1}$), whereas soils of 69.1 and 15.9 per cent area are medium (10-20 mg $kg^{-1}$) and high (>20 mg $kg^{-1}$) in available sulphur content, respectively.

Soils we analysed for available (DTPA extractable) micronutrients (Zn, Cu, Fe and Mn) and found that, all the soils are sufficient in available iron and manganese whereas 9.4 and 2.9 per cent area are deficient in available zinc and copper, respectively. Soils of 24.6 per cent area of block is deficient (<0.50 mg $kg^{-1}$) in available boron ( hot water soluble ).

## Recommendations

1. More than 50% soils of Kundahit Block pose major challenges in crop production due to high soil acidity. This need to be addressed first before resorting to fertilizer use for higher farm profits. Farmers should be encouraged to ameliorate soil acidity problem to sustain soil health and high crop yields.
2. Soils of all the blocks are of poor fertility as evidenced by low organic matter level and high Phosphate deficiency (caused by high soil acidity and P fixation). Potassium deficiency in these soils is widespread (mainly due to its non-application and high crop removal). Application of fertilizers and manures as per requirement of crop (s)/ cropping sequence based upon soil testing in a site specific mode, is therefore advocated.
3. Zinc is deficient in about 10% soils of the block (lowlands with continuous rice cropping). Application of 5kg Zn/ha at transplanting and/or 2 to 3 sprayings of 0.5% $ZnSO_4$ solution in standing crop will help in improving the situation. Boron (24.6%) and Sulphur (13.7%) deficiency are also found in the soils. Farmers need to apply these nutrients based on soil tests.

## References

Berger, K. C. and Truog, E. 1940. *J. Am.* Soc. *Agron.* 32:297

FAI, 2011 – 2012. Fertiliser and Agriculture Statistics, Eastern Region.

Follet, R. H. and Lindsay , W. L. 1970. Tech. Bull. Colo. Agric. Exp. Station 110.

Haldar, A. K., Srivastava, R., Thampi, C. J., Sarkar, D., Singh, D. S., Sehgal, J and Velayutham, M. 1996. Soils of Bihar for optimizing land use. NBSS Publ. 50b. (Soils of India Series), National Bureau of Soil Survey and Land Use Planning, Nagpur, India, pp. 70+4 sheets soil Map (1:500,000 scale).

Hatcher, J. T. and Wilcox, L. V. 1950. *Analyt. Chem.* 22: 567.

I.A.R.I. 1970. Soil survey manual, All India Soil and Land Use Organization, Indian Agricultural Research Institute, New Delhi.

Lindsay, W. L. & Norvell, W.A. 1978. Development of a DTPA micronutrients soil test for Zn, Fe, Mn and Cu. *Soil Sci. Soc. Am. Proceedings*: 42: 421-428.

Mehta, V. S., Singh, V and Singh, R. P. 1988. *J. Indian Society of Soil Science*, 36: 743.

Mishra, R. K. 2004. Planning for Food and Nutritional Security in Jharkhand, Published by Agricultural Data Bank, BAU, Ranchi, Jharkhand, p. 275

NBSS & LUP 2006. Assessment and mapping of some important soil parameters including soil acidity for the State of Jharkhand (1:50,000 scale) towards rational land use plan, NBSS & LUP Report No.946, National Bureau of Soil Survey and Land Use Planning, Nagpur, India, pp.243.

Page, A. L., Miller, R. H. and Keeney, D. R. 1982. Method of Soil Analysis, Part-II, Chemical and Microbiological Properties, Soil Sci. Soc. Am. & Am. Soc. Agron. Madison, Wisconsin, USA.

Sahoo, A,K.; Singh, S.k,; Sarkar Dipak; Sarkar, A.K; Agarwal, B.K.; Das, T.H.; Nayak, D.C.; Mukhopadhyay, S. and Banerjee, T. 2013. Mapping of Soil Acidity and Nutrient Status at Block Level of Hazaribagh District, Jharkhand, NBSS Report No. 1052C, NBSS & LUP (ICAR), Nagpur, 155p.

Sen, P.,Singh, S.K., Sarkar, D. and Majumdar, A. 2012. Farmer's Advisory Services for Fertilizer and Crop Planning in the State of West Bengal- A Novel Approach to reach out to the farmers. *Fertilizer Marketing News* 43: 1-7.

Singh Dhyan, Chhonkar, P. K. and Pandey, R. N. 2004. Soil Plant and Water Analysis, A Manual, IARI, New Delhi.

Tandon, H. L. S. (Ed) 1999. Methods of analysis of soils, plants, waters and fertilizers. Fertilizer Development and Consultation Organisation, New Delhi, India.

William, C. H. and Stainbergs, A. 1959. *Aust. J. Agric. Res.* 10: 342

www.census2011.co.in. Census of Jharkhand State in Census of India, 2011.

❑❑❑

CHAPTER - 9

# Task Ahead

**A.K. Sarkar**

## 1. Why Plant Nutrient Disorders?

Plant Nutrient Disorders result from nutrient imbalances in soils and plants. In-adequate use of one or more nutrients can result in nutrient imbalances. Presence or use of a nutrient in excess amount can hinder the absorption or uptake of another nutrient and thus,cause nutrient disorders. Lime induced iron chlorosis or use of excess Potassium can hinder the absorption of $Ca^{2+}$/and, or $Mg^{2+}$ by plants. Enhanced level of Mg can decrease the uptake of Zn. Excessive use of NPK fertilisers can result in micronutrient deficiencies.

Soils are being regularly depleted of their native reserves, due to intensive cropping and non-application of essential nutrients (primary, secondary and micronutrients). Lack or restricted use of quality organic manures, such as compost, green manures, crop residues etc. and cultivation of HYVs in most crops have accentuated the problem.

Occurrence of plant nutrient disorders, in fact, originate from a depleted soil fertility level. This results from mining of soil nutrients and their non-replenishment from external sources. Soil properties, such as pH, organic carbon, texture, CEC, nature and content of clay minerals govern the 'available nutrient pool' of soils. This is modified to a considerable extent by factors, such as, soil moisture, temperature and climatic aberrations. These bring about changes in the activity of soil microorganisms and affects the mineralization-immobilization processes. Nature of crops (shallow-rooted or deep-rooted ) also influence the amount of nutrients absorbed or taken up from soils.

## 2. Are Plant Nutrient Disorders, Site or Crop Specific?

Nutrient disorders are site or crop specific. High phosphate fixation in low pH soils, result in deficient soil available P status in Black soils (Vertisols). High clay content also poses problems of Zn availability. Zn deficiency is also common in high pH and calcareous soils. Boron deficiency is common in red and lateritic, acidic, coarse-textured alluvial and highly calcareous soils. Fe deficiency is common in coarse textured, high pH soils with high $CaCO_3$ and organic matter contents.

Crops, in general, show chlorotic appearance, poor and stunted growth, less tillering, weak stems, poor flowering, as well as poor grain formation resulting in poor crop yields in case of nutrient disorders.

Thus, it is essential to quantify the capacity of soil to supply nutrients before planting of crops. This is because of the fact that, by the time plants exhibit deficiency symptoms, a reduction in yield potential has already taken place. Examples of crop specificity for a particular nutrient is evident in indicator plants for nutrient deficiencies. Maize or citrus is an indicator of Zn deficiency, cauliflower for boron, rice or sorghum for iron, brassica or cauliflower for molybdenum and groundnut for sulphur, etc.

### 3. Is Knowledge of Plant Nutrient uptake Pattern Essential?

Information and knowledge of nutrient uptake trends of a crop, can provide solutions to check plant nutrient disorders through manipulation of soil-crop environment. Rattan and Goswami (2009) indicated that, there are wide variations in nutrient uptake by crops in soils, but some generalizations can be made based on research data.

- Uptake of N is equal to that of K.
- Uptake of Ca is half of N and K uptake.
- Uptake of P,Mg,& S are more or less similar.
- Uptake of Fe is about 1/100 of N and K uptake.
- Uptake of Mn,Zn,B,is about ½ of Fe uptake.
- Uptake of Cu is about ½ of Zn,B,Mn.
- Uptake of Mo is about $1/10^{th}$ of Zn.

These are approximations or trends in the uptake of nutrients. But, the crops show wide variations in this regard. LTFE results show that, among cereals, nutrient uptake is highest in fingermillet. Rice and wheat crops consume similar quantities of N. Uptake of K is similar to that of N in rice, but less than that in wheat. P uptake is also more in rice than that in wheat. Generally, N uptake by leguminous crops is higher, while K uptake is more in oilseed crops, compared to legumes. S and Ca removal is high in oilseed crops compared to legumes. It has been observed that, soybean, being both a legume as well as oilseed crop, removes larger quantities of N,P,K,Ca,Mg&S.

Generally, cereals, pulses and oilseed crops require 1-6,5-13, and 5-20 kg S/t of economic produce with mean values of 4,8, and 12 kgS/t. Forage crops remove larger quantities of N,P, and K compared to root, fruit and plantation crops. Fruit and root crops remove more K than N, while in forages, the order is reverse. Forages also remove more P than other crops.

## 4. What are the Major Concerns that Need to be Addressed?

Major concerns are:

- Soil organic matter depletion
- Imbalanced nutrient use
- Checking soil erosion/land degradation
- Making soil testing effective
- Improving nutrient use efficiency

**Soil organic matter depletion:** Integrated use of plant nutrient sources, such as chemical fertilisers, organic manures, crop residues, green manures, bio-fertilisers etc. as per site specific availability need to be promoted on a large scale. Lack of this approach has resulted in deteriorating soil health. Long term studies in different cropping systems have shown that, combined use of 5 to 10 t/ha of compost with balanced NPK application as per crop requirements, improves the physical, chemical and biological properties of soils and promotes sustainable crop production. This practice reduces the dose of NPK application and increase profits. We have to improve the quality of organic manures to reduce the rate of their application in soils. This will help in reducing the carrying cost of these bulky manures from the production site to the fields of farmers. Crop yields and quality of produce, improves with organic manures, along with a substantial improvement in the keeping quality. Farmers need to be encouraged to adopt such practices through practical training on compost preparation and their method of application in fields.

**Imbalanced nutrient use:** Disproportionate nutrient use (use of N in excess, poor or restricted use of K and P and low or no use of S, Zn, or B in soils) is a threat to long term sustenance of soil fertility. Both, yield of crops and quality of harvest and the ability of crops/plants to withstand stress, is negatively influenced by wider N,P,K, ratios. Farmers perspective is high production at reasonable cost. In high production agriculture, nutrient interactions play a significant role in raising farm profits. Nutrient elements, like K, S, or B improves quality of crops and their absence from the fertilization schedule, lowers their market value. Kanwar (2009) noted that nutrient imbalance is most conspicuous in northern region of India, especially in the states of Punjab and Haryana, which use 200-250 kg NPK/ha. This (absence of secondary and micronutrients) has resulted in lower nutrient use efficiency, greater loss of nitrogen and increased deterioration of groundwater and soil quality with stagnation of crop yields. This is undesirable and can not be allowed to continue.

**Checking soil erosion/land degradation:** Adequate care must be taken to conserve soil and water resources. This has been dealt in a recent publication (Sarkar, 2014). Checking runoff in hilly areas and high rainfall regions, minimizes loss of top soil (thereby, losing soil organic matter & plant nutrients), conserve soil fertility. Soil erosion and degradation can be reduced to a considerable extent, by promoting land use as per its capability and resorting to remedial measures as per need.

Problem soils (acid, waterlogged, saline, alkali etc.) need to be ameliorated/ reclaimed before resorting to chemical fertilisation. For example, extensive research on managing acid soils (Sarkar,2013) has conclusively shown that, application of 50 to 75% of recommended NPK + cmpost+2 to 4 q lime/ha to crops can double the yield of crops in acid soil regions with a positive impact on the soil fertility status. Long term experiments in acid soils with pulse, oilseed or maize based crop sequences have substantiated these observations. Such studies have shown that fertiliser application in crops without amelioration of soil acidity is in-effective and should not be adopted.

**Making soil testing effective:** Soil testing continues to be the most widely adopted method for soil fertility evaluation. Fertiliser application based on soil tests are effective. What is required to identify the limitations of this practice and remove them.

Performance of soil test laboratories at district level is poor in most cases. This is due to lack of qualified personnels, poor instrumentation and infrastructure, lack of coordination with farmers, poor/incorrect soil test data interpretation and complete absence of any follow-up. Most laboratories still do not have facilities to diagnose secondary and micronutrient deficiencies. Soil sampling in most cases is not based on GIS. Most STLs do not prepare soil health cards for farmers. These deficiencies need to be corrected, to improve the acceptance of soil testing among farmers. Public-private partnership can help in making this effective. Dwivedi (2014) identified the research gaps in soil testing. These are i) lack of soil sampling norms (not GIS/GPS/Grid based) ii) problems associated with refinement of soil testing methods especially for N, S, B.Fe., and iii) incomplete and inadequate fertilizer recommendation related to soil test based fertility rating, mostly confined to N, P, and K recommendation, and not for different yield levels and micronutrients based on critical levels.

Keeping in view, the fast changing scenario of soil, water, crop, and environment and their impact on crop production, there is need to establish

Referral laboratories in different states to monitor nutrient supply in the soil-plant system in an integrated manner. Besides soil tests, plant tests are useful for fruit crops, forest trees and perennial crops for making mid-term corrections on nutrient use (deficiency/or luxury consumption) and improve crop performances. Similarly, poor water quality affects plant growth adversely due to the presence of soluble salts or toxic substances. Such water resources need treatments before use, for better results. Environmental changes (elevated $CO_2$, temperature, moisture and their interactions) greatly influence soil processes and plant metabolism, and also soil carbon storage. Such an approach for integration is required to monitor intensive agriculture vis-à-vis soil quality.

**Improving Nutrient Use Efficiency:** This is a major challenge , though opportunities exist. Normally, at lower yield targets efficiency is higher. But, we need an increase in the nutrient use efficiency (NUE) at higher yield level, so that, it is economically viable. Right time, rate, and placement of fertilisers improves NUE. Over-and under-application of nutrients result in lower nutrient use efficiencies, losses in crop yields, and deterioration in soil health. Soil test based use of nutrients, help in maintaining a positive soil nutrient balance. Similarly, nutrient supply in tune with crop demand, help in minimizing losses by fixation, volatilization, leaching and other processes. Plant tissue tests, as a diagnostic tool, is useful to assess nutrient status of growing plants. Chlorophyll meters, for example, can help in minimizing nitrogen starvation by plants by supplementation as per need. Leaf colour charts or use of on- the- go N-sensors, can help optimize N application in crops.

Nutrient placements help in minimizing fixation or reversion of soluble nutrients to insoluble forms. For nitrogen application in submerged and aerobic soils, one has to consider a sub-surface or surface placement of fertilizers for higher NUE. Similarly, band placement of P fertilizers can help in reduced soil-fertiliser P contact and improve P utilization by crops. Soil vs. foliar application of micronutrients, or use of chelated sources, help in increased utilization of such nutrients. Farmers need to be made aware of such methodologies to economise nutrient use for crop production.

## 5. How Best, We can Manage Plant Nutrient Disorders?

Plant nutrient disorders can be effectively managed by taking a composite view of the soil-water-plant-climate-farmer inter-relationships.

The productivity of problem soils, such as acid, alkali or saline soils can be improved by adopting suitable amelioration/reclamation measures. Application of lime/dolomite/paper mill sludge/basic slag (as Ca source) in furrows and addition of recommended nutrients will increase acid soil productivity. Leaching of soluble salts in saline soils and reclamation of alkali soils by adding gypsum will increase productivity of salt-affected soils.

In rain-fed areas, intercropping of pulses (pigeon-pea/black-gram) in maize or rice- based crop sequences, will increase the system productivity, keep the soil covered for longer periods and help conserve soil and water. Covering the soil with legumes, grasses, mulches, and residues help protect soil erosion, moderate soil temperature, reduce evaporation, and enhance soil biological activity.

Climate change has the risk of depleting the soil organic carbon stocks. Atmospheric carbon maintains a balance with soil carbon through trees, plants and soils. Therefore, practices such as deforestation, intensive cropping, need to be monitored regularly, for their effect on soil organic carbon. Crop cover, minimum tillage, climate, nature of soil, and management practices help soils to store or sequester carbon, which will help in sustaining soil fertility. Use of cattle manures and wetland rice cultivation are responsible for methane emissions. System of rice intensification (SRI) and alternate wetting and drying/ partial submergence/aerobic-rice technology can help reduce $CH_4$ emission.

A major problem in diagnosis of nutrient deficiency/sufficiency, is the small land holding and variation in management practices of individual farmers. Farmers need to be trained on the new and emerging technologies for wider adoption. 'Soil health cards', 'soil fertility maps' need to be a regular and popular practice among farmers and extension workers. Soil tests or plant tissue tests will continue to be the primary method of nutrient deficiency diagnosis. But, in well-managed farms, irrigated agriculture, precision agriculture and under high input agriculture, new diagnostic tools, such as, site specific nutrient management, nutrient expert system, or use of sensor based techniques need to be put in place for accuracy, better results & higher farm profits.

Targetted yield concept was an important development in soil test based fertiliser recommendation in crops. This provides the nutrient needs of crops with different yield targets, per cent contribution of soil available nutrient and that of the applied fertilisers. In recent years, SSNM has been successfully carried out in a variety of crops (Srivastava *et al.* 2014). This helps in nutrient application at optimal rate and time to achieve high yields with greater use efficiency of added input (Dobermann *et. al.* 2003a , 2003b). There are five steps through which SSNM is accomplished. These are:

- Establishment of a yield target.
- Estimation of actual yield response to N, P, K fertilisers.
- Selection of NPK rates based on expected yield response to fertiliser application, considering agronomic effectiveness and nutrient balances.

- Application of fertilisers to meet the crop demand for nutrients at critical growth stages.
- Optimization of nutrient use efficiencies ( Hach& Tan, 2007).

Based on SSNM principles, a decision support system known as 'Nutrient Expert' was developed for Maize in 2010. This is an easy to use, interactive, computer-based decision tool, that can rapidly provide nutrient recommendations for individual farmers fields. Use of sensors in soil and water testing is a possibility, especially for precision agriculture. Sensor based techniques can help in sensing nutrients and moisture in soils for yield and quality of export oriented crops/ precision farming and facilitate simultaneous application of nutrients at right amount and right place for improved use efficiency and cost effectiveness.

What are the suggested steps in Plant Nutrient Disorder Diagnosis.

**STEP I Detection**

- Site history and characteristics (soil fertility maps/other informations)
- Plant growth pattern
- Deficiency or toxicity symptoms ( visual appearance).

**STEP II Diagnosis**

- Soil analysis
- Plant analysis
- Water analysis
- Environmental characters (as per need)
- Report from soil test lab
- Refer to plant clinic (if needed)
- Refer to referral lab (if required)
- Consultation, diagnosis and recommendation.

**STEP III Management**

- Farmer's views/options/reactions/preferences
- Soil application of deficient nutrients
- Foliar sprays of deficient nutrients (as per need)
- Address problems related to soil moisture/tillage/aeration/soil organic matter

- Use of amendments (if necessary )
- Use of mulches/residues (if necessary)
- Training of farmers/farmers groups/issue of soil health cards/follow-up
- Documentation/IT, ICT based dissemination among farmers, technical staff.

### STEP IV Impact Analysis

- Analysis of yield and quality improvements
- Analysis of farmers feed back & knowledge gain
- Analysis of economic gain
- Analysis of nutrition and health issues (long term impact)
- Analysis of confidence build-up among village communities
- Assessment and follow-up.

## References

Dobermann, A. *et al.* 2003a. Soil fertility and indigenous supply in irrigated rice domains of Asia. *Agron.* J. 95:913-23.

Dobermann, A. *et al.* 2003. Estimating indigenous nutrient supplies for site specific nutrient management in irrigated rice. *Agron.* J. 95:924-35.

Dwivedi, B. S. 2014. In 41st. R. V. Tamhane Memorial Lecture on 'Revisiting soil testing and fertiliser use research'. Indian Society of Soil Science, New Delhi.

Hach, C.V. and Tan, P. S. 2007. Study on Site Specific Nutrient Management (SSNM) for high yielding rice in the Mekong Delta. *Omonrice* 15:144-152.

Kanwar, J.S. 2009. In' Fundamentals of Soil Science'. Indian Soc. of Soil Science Publ. New Delhi.

Rattan, R.K. and Goswami, N. N. 2009. In 'Fundamentals of Soil Science'. Indian Soc. of Soil Science Publ., New Delhi.

Sarkar, A. K. 2013. In 'Acid soils-their chemistry and management'. New India Publ. Agency. New Delhi-34.

Sarkar, A.K. 2014. In 'Management of soil and water resources for sustainable agricultural production' Agrotech Publ. Agency, Udaipur, Rajasthan.

Srivastava, A. K., Das, S.N., Malhotra, S.K., & Majumdar, Kaushik 2014. SSNM-based rationale of fertiliser use in perennial crops: A review. *Indian J. Agril. Sci.* 84(1) : 3-17.

❑❑❑

FSC
www.fsc.org
MIX
Papier | Fördert
gute Waldnutzung
FSC® C083411

Zeitfracht Medien GmbH
Ferdinand-Jühlke-Straße 7
99095 Erfurt, Deutschland
produktsicherheit@kolibri360.de